W9-BGC-426

LABORATORY EXPERIMENTS

Introduction to
CHEMISTRY

7th Edition

T. R. Dickson
Cabrillo College

JOHN WILEY & SONS, INC.
New York   Chichester   Brisbane   Toronto   Singapore

Copyright © 1995 by John Wiley & Sons, Inc.

All rights reserved.

Reproduction or translation of any part of this work
beyond that permitted by Sections 107 and 108 of
the 1976 United States Copyright Act without the
permission of the copyright owner is unlawful.
Requests for permission or further information
should be addressed to the Permissions Department,
John Wiley & Sons, Inc.

ISBN 0-471-05879-3

Printed in the United States of America

10 9 8 7 6 5 4 3 2

Preface

Chemistry is an experimental science and laboratory experience is a necessary part of the study of chemistry. Many important concepts are vitalized in laboratory exercises. Furthermore, much nonverbal learning is accomplished by the student through laboratory work. This laboratory manual is designed for a first course in chemistry. It is intended for students who have had no laboratory experience. Along with vitalizing important chemical concepts this manual emphasizes laboratory techniques, procedures, and safety.

A variety of experiments is included so an instructor can select the preferred laboratory sequence. Some experiments stress observations of the properties and reactions of chemicals. Others emphasize the collection of data and the use of the data in calculations. Students are provided with detailed directions but are also challenged by situations in which they must draw conclusions, describe their observations, identify unknowns, and carry out calculations.

The seventh edition of this manual has been revised using the experience of students and the advice of instructors who have used it. An effort has been made to cut or reduce the use of potentially hazardous and toxic chemicals. Acids, bases and hydrocarbons are the only hazardous chemicals used. (Acetic anhydride is required in the preparation of aspirin experiment.) The proper handling of any hazardous chemicals is emphasized by caution statements throughout the manual. None of the experiments require special or elaborate equipment or chemicals. Nevertheless, the experiments do illustrate the basic ideas of chemistry and important laboratory techniques.

The first few pages of this manual include safety rules, a laboratory equipment list and a note to students. Appendix 1 describes general laboratory techniques. These techniques are referenced in various experiments so that students can learn about them as needed. They can be used as the basis of laboratory lectures. A synopsis of the various experiments follows.

I would like to thank the students and instructors who helped me prepare the seventh edition of this manual.

T. R. Dickson

Aptos, California   1994

Table of Contents

Laboratory Rules and Safety

Typical Laboratory Desk Equipment

Typical Laboratory Equipment

Note to Students

LABORATORY RULES AND SAFETY

1. Eye protection must be worn when you are working in the laboratory.

2. Follow the instructions given in an experiment. Never do unauthorized experiments.

3. Read all labels before using chemicals. Make sure you have the correct chemical.

4. When heating a liquid in a test tube or carrying out a reaction in a test tube, never point the mouth of the tube at yourself or your neighbor.

5. Never taste a laboratory chemical. Avoid breathing gases given off in reactions.

6. If dangerous gases are given off during an experiment, do the experiment in a fume hood as instructed.

7. Report any injury to the instructor at once, no matter how slight it may appear.

8. Never pour water into concentrated acid. Always slowly pour the acid into water.

9. Return all waste chemicals to the containers indicated by your instructor. If no containers are required, pour waste liquids directly down the drain and then run water down the drain. All waste solids (except those so indicated) should be placed in waste containers. Do not throw solids in the sink. When in doubt, ask about the disposal of wastes.

10. Always moisten or lubricate glass tubing when it is being connected to rubber tubing or inserted in rubber stoppers.

11. If you spill any solid chemicals, clean them up. If you spill any liquid chemicals, clean them up with paper towel. If you spill any acids or bases on the desk or floor, sprinkle some sodium hydrogen carbonate (baking soda) on the spill to neutralize it, then wipe it up.

12. If you splash any acid or base in your eyes, face, hands, or clothing flush with plenty of water. Keep the nearest source of water in mind so you can go to it as quickly as possible. If a splash occurs, notify your instructor who will recommend any further action.

13. No eating, drinking, or smoking in the laboratory.

14. Do not place your books or papers on any laboratory chemicals. Clean your laboratory desk before and after use.

15. Keep your locker clean and wash any desk equipment after you use it. Always wash your hands before leaving the laboratory.

TYPICAL LABORATORY DESK EQUIPMENT

2 Beakers, 100 mL

1 Beaker, 250 mL

1 Beaker, 400 mL

1 Crucible, porcelain

1 Crucible cover

1 Graduated cylinder, 10 mL

1 Graduated cylinder, 100 mL

1 Evaporating dish

2 Erlenmeyer flasks, 50 mL

2 Erlenmeyer flasks, 250 mL

1 Funnel

1 Test tube rack

1 Thermometer, 110 °C

1 Crucible tongs

1 Tweezers

1 Test tube holder

1 Test tube brush

6 Test tubes, 1.8 cm x 15 cm (large)

6 Test tubes, 1.3 cm x 10 cm (small)

2 Watch glasses

1 Wire gauze

1 Clay triangle

TYPICAL LABORATORY EQUIPMENT

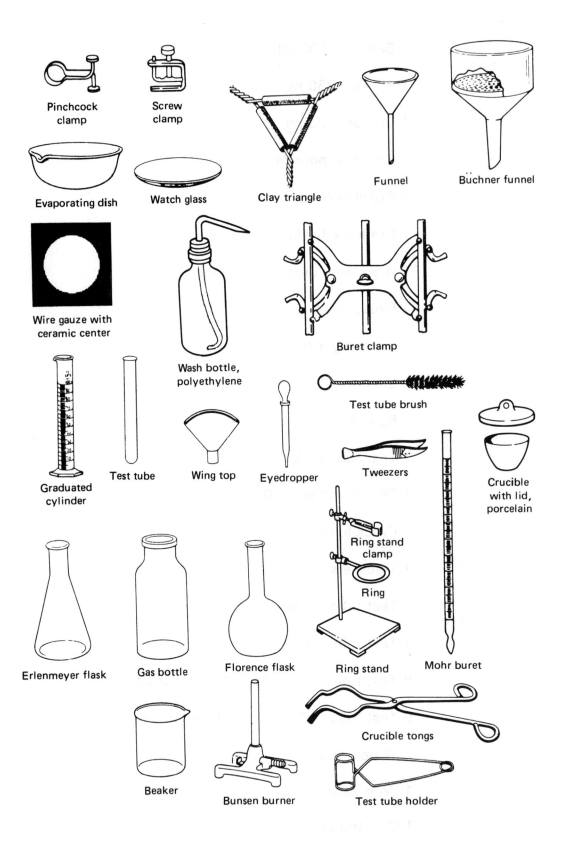

Pinchcock clamp

Screw clamp

Clay triangle

Funnel

Buchner funnel

Evaporating dish

Watch glass

Wire gauze with ceramic center

Wash bottle, polyethylene

Buret clamp

Graduated cylinder

Test tube

Wing top

Eyedropper

Test tube brush

Tweezers

Crucible with lid, porcelain

Erlenmeyer flask

Gas bottle

Florence flask

Ring stand clamp

Ring

Ring stand

Mohr buret

Beaker

Bunsen burner

Crucible tongs

Test tube holder

Note to Students

This laboratory book contains a variety of chemistry experiences that relate directly to the concepts and ideas that are part of your study of chemistry. Each experiment has specific objectives. The overall objectives of laboratory work are the development of your skills in various laboratory techniques, practice in making observations and the collection and use of quantitative experimental data. The following topics are important to your laboratory process and experience.

LABORATORY NOTEBOOK

It is useful to keep a laboratory notebook to record:
information and comments,
laboratory lecture notes,
summaries of laboratory experiments and,
especially, laboratory observations and experimental data.

Here are some guidelines for keeping a laboratory notebook.

1. Use a bound notebook rather than a loose-leaf notebook.

2. Make a title page with your name, course information and other pertinent information.

3. Leave a few pages at the front for a possible table of contents.

4. If the pages are not numbered, you should number them in ink.

5. Each time you enter information or data for an experiment the first entry page should have the current date and a reference indicating the experiment by title or number.

6. Your notes and comments in the notebook can be done in pencil.

7. All experimental observations and data should be recorded in ink.

8. If you make an error in the notebook do not erase it. Just draw a single line through the error and then add the corrected data or observation. Never obliterate, erase or scratch out an error.

9. Never remove or tear pages from the notebook. Pages can be skipped.

10. Be sure to label information in the notebook so that you can refer to it later.

11. Use the observations and data from your notebook to complete your laboratory report sheet.

PREPARING FOR A LABORATORY ASSIGNMENT

To prepare for a laboratory assignment you need to carefully read the objectives, terms to know and discussion sections of the exercise. Be familiar with the terms to know in each exercise so that you can follow or participate in any discussion of the concepts, techniques and equipment involved in the experiment. Prepare an entry in your laboratory notebook for the assigned experiment and add any notes or comments to the notebook. Review the laboratory procedure and list any unique equipment and materials needed. If an experiment or part of an experiment includes several steps involved in the collection of quantitative data make a flow chart of the process. A flow chart serves as a guide to the steps in the experiment and any required measurements.

As an example of a flow chart consider the laboratory procedure for Experiment 1, Popcorn: Water in a Mixture.

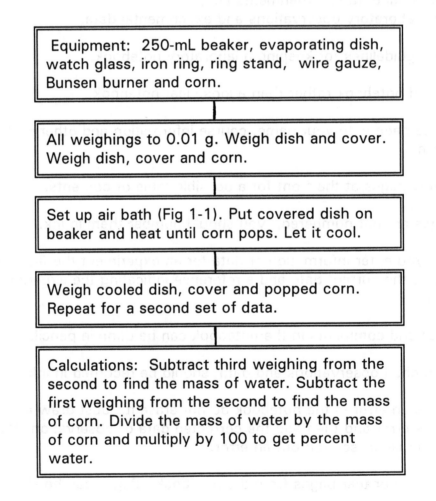

Equipment: 250-mL beaker, evaporating dish, watch glass, iron ring, ring stand, wire gauze, Bunsen burner and corn.

All weighings to 0.01 g. Weigh dish and cover. Weigh dish, cover and corn.

Set up air bath (Fig 1-1). Put covered dish on beaker and heat until corn pops. Let it cool.

Weigh cooled dish, cover and popped corn. Repeat for a second set of data.

Calculations: Subtract third weighing from the second to find the mass of water. Subtract the first weighing from the second to find the mass of corn. Divide the mass of water by the mass of corn and multiply by 100 to get percent water.

With the appropriate prelaboratory preparation your laboratory experience will be more interesting and rewarding. Enjoy your experiences and Bon Cognitio!

1 Popcorn: Water in a Food

Objective

To determine the percent-by-mass of water in popcorn.

Terms to Know

Percent-by-Mass - The number of grams of a part in 100 parts of the whole. The percent-by-mass of a component is determined by dividing the mass of a component by the mass of the whole and multiplying by 100.

Evaporating Dish - A bowl shaped porcelain vessel used to contain chemicals so that they can be heated.

Watch Glass - A round and concave piece of glass which can be used to cover an evaporating dish.

Discussion

Foods are composed mainly of chemicals called carbohydrates, fats and proteins along with small amounts of vitamins, minerals and variable amounts of water. Popcorn kernels are corn seeds consisting of a variety of compounds encased in a hard shell. The main component of the mixture is the carbohydrate starch. The starch is in the form of granules that are impregnated with water. This experiment is designed to find the percent-by-mass of water in popcorn kernels.

To determine the percent-by-mass of water in popcorn, the kernels are popped and dried. First, a sample of kernels is weighed. The sample is heated to pop the kernels and to dry the popped corn. The popped corn is then weighed. The mass of water driven off can be found by subtracting the mass of the popped corn from the mass of the unpopped corn. The percent of water is found by dividing the mass of water by the mass of the sample of kernels and multiplying by 100. A kernel which does not pop (a "dud") often has a crack in the shell but it will still lose water and dry out.

1

Laboratory Procedure

For this experiment you will need a 250-mL beaker, an evaporating dish and a watch glass cover for the dish. Set up an iron ring supported by a ring stand as shown in Fig. 1-1. Place a wire gauze on the ring to support the beaker. See Part 1 of Appendix 1 for a discussion of using laboratory balances and weighing materials. Obtain 15 kernels of unpopped corn on a clean piece of paper or in a small beaker. Weigh the evaporating dish and watch glass cover to the nearest 0.01 g. Place the kernels in the dish, cover it and weigh to the nearest 0.01 g.

See Part 2 of Appendix 2 for a discussion of the Bunsen burner. Position a Bunsen burner under the ring supported by the ring stand and adjust the level of the ring so that the flame will reach the gauze. (See Figure 1-1.) Do not light the burner. Place the 250-mL beaker on the gauze and place the covered evaporating dish on top of the beaker as shown in the figure. The beaker will serve as an air bath to heat the evaporating dish. This avoids direct contact of the flame and the dish.

Light the burner and make sure that the hot part of the flame is in contact with the bottom of the beaker. Allow the system to heat until the kernels pop. As soon as they all pop, shut off the heat. Too much heating may cause the popcorn to scorch or burn. If this happens, let the apparatus cool and start over with a new sample of kernels. If some of the kernels are "duds", just continue with the experiment.

Allow the dish to cool to room temperature. When it is cool enough to touch, you can move the dish to the top of the desk so that it will cool faster. After the covered dish has reached room temperature, weigh it to 0.01 g. Do not try to weigh a hot object. Let it cool to room temperature. Clean the dish and repeat the experiment to obtain a second set of data.

The mass of water driven off from the popped corn can be found by subtracting the mass of the dish with the popped corn from the mass of the dish with the unpopped corn. Determine the mass of water. Find the mass of the unpopped kernels by subtracting the mass of the covered dish from the mass of the dish and the unpopped corn. Use the data to calculate the percent-by-mass of water in the popcorn by dividing the mass of water by the mass of the unpopped kernels and multiplying by 100. After calculating the percent for each sample, calculate the average of your two results. If your two results seem too far apart, consult your instructor.

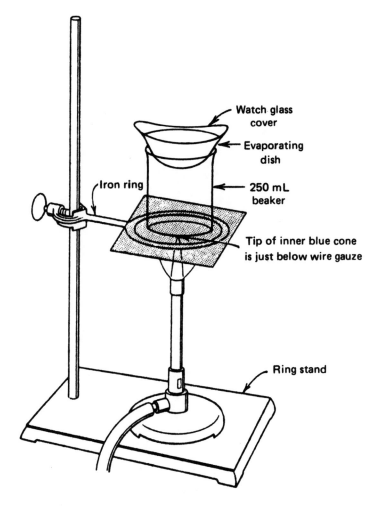

Fig. 1-1 Apparatus for Air Bath

2 Density

Objective

The purpose of this experiment is to investigate the topic of density by determining the densities of some materials.

Terms to Know

Density - The mass per unit volume of a material.   D = m/V

Property Factor - A type of conversion factor that relates two properties to one another. Density serves as a property factor that relates the mass of a material to the volume of the material.

Graduated Cylinder - A calibrated piece of glassware used to measure precise volumes of liquids.

Discussion

There is a lesson in the old bromide: "Which is heavier, a pound of feathers or a pound of lead?" Neither is heavier since a pound of anything is still a pound. It is not possible to compare masses of objects unless it is done on the basis of the same portion of each object. For instance, it is possible to say that one person is heavier than another person since the comparison is on a pound per person basis. Since all materials have mass and occupy space, it is possible to compare materials by stating the amount of mass contained in a specific volume of each material. Density is defined as the mass per unit volume of a material, where mass is usually expressed in grams and volume in cubic centimeters or milliliters. Density can be determined by measuring the mass and volume of a sample of a material and then dividing the mass by the volume:

$$\textbf{Density} = \frac{\textbf{mass}}{\textbf{volume}} \quad \text{or} \quad D = \frac{m}{V}$$

8

The density of any material is a characteristic property of the material and can help identify the material. Furthermore, density can be used to relate the mass of a sample of the material to the volume or the volume to the mass. That is, the density can be used as a property conversion factor to find the volume given the mass,

$$V = m \frac{1}{D}$$

or to find the mass given the volume,

$$m = VD$$

In this experiment, the densities of several materials are determined and densities are used as conversion factors.

The volumes of liquids are often determined in the laboratory by use of graduated cylinders. These cylinders are calibrated in the factory so that they will contain certain specific volumes. Lines are etched on the outside surface of the cylinder to indicate the positions corresponding to various volumes. Distances between subsequent lines on the cylinder correspond to specific volumes. Volumes of liquids can be measured by pouring a liquid sample into the cylinder and observing the position of the surface of the liquid. Some liquids, such as water, do not form a flat surface when placed in a glass cylinder, but instead they form a concave surface called a meniscus. Consequently, glass graduated cylinders are calibrated so that the volumes can be read by observing the lowest portion of the curved meniscus. In other words, read the bottom of the meniscus. The volume of the liquid can be measured by observing the position of the meniscus with respect to the calibration lines on the cylinder. Consider an example of the reading of the volume of a liquid in a graduated cylinder. In Fig. 2-1b, the position of meniscus coincides with the 15-mL mark and the volume of the

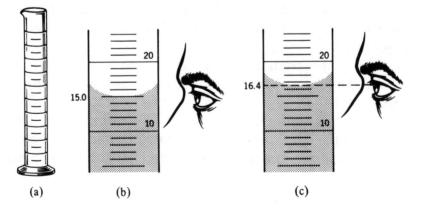

(a)   (b)   (c)

Fig. 2-1

liquid would be read as 15.0 mL. However, in Fig. 2-1c the meniscus level does not coincide with any particular line but falls between lines. The estimation of the position of the meniscus in this case illustrates a very important method involved in the measurement: interpolation. See the discussion at the beginning of Appendix 1. Interpolation can be carried out by visually splitting the distance between two lines into equal parts and then approximating the position at which the meniscus lies between the two lines. In the figure, if the distance between lines is split into 10 parts then the meniscus appears to lie at the 16.4-mL position. The interpolation method is very important in volume measurements. The number of digits to be read by interpolation depends on how the volume measuring device is calibrated. When in doubt about how to read a cylinder consult your instructor.

Laboratory Procedure

Measure all masses to the nearest 0.01 g. Use of balances and methods of weighing materials are discussed in Part 1 of Appendix 1. Be careful to use the balances correctly. Your instructor will demonstrate the proper use of the laboratory balances. Balances are delicate instruments, so use them with care and respect. It is a good idea to use the same balance for subsequent weighings. The volumes of materials will be determined using graduated cylinders. The use of graduated cylinders is discussed in Part 3 of Appendix 1. Since the number of significant digits in the volumes will dictate the number of digits in your calculated densities, be sure to read the volumes carefully.

1. DENSITY OF WATER

Dry your 10-mL graduated cylinder and find its mass to the nearest 0.01 g. Pour about 9 mL of water into the cylinder and carefully read the volume to the nearest 0.1 mL. Now weigh the cylinder plus the water to the nearest 0.01 g. Calculate the mass of the water sample and use the mass and volume to find the density of the water.

2. DENSITY OF AN UNKNOWN LIQUID

Thoroughly dry your 10-mL graduated cylinder and weigh it to the nearest 0.01 g. Pour about 9 mL of unknown liquid into the cylinder and carefully read the volume to the nearest 0.1 mL. Now weigh the cylinder plus the liquid to the nearest 0.01 g. Calculate the mass of the liquid sample and use the data to find the density of the liquid.

3. DENSITY OF SOLIDS

The density of a solid with an irregular shape can be found by first determining the mass of a sample and then placing the sample in a graduated cylinder partly filled with water (or some liquid in which the solid does not float or dissolve). The solid will displace a volume of liquid equal to its volume. Thus, by noting the position of the meniscus before and after the addition of the solid, the volume of the solid can be determined. (See Fig. 2-2.)

Fig. 2-2 Determination of the volume of a solid.

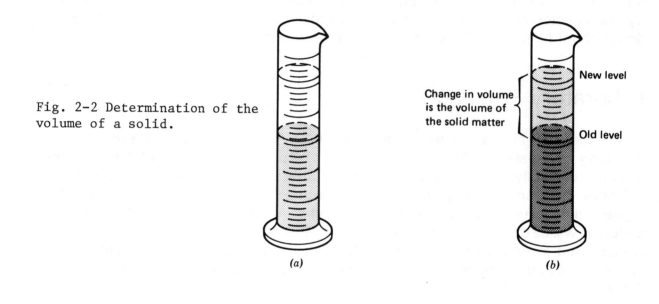

Change in volume is the volume of the solid matter

New level

Old level

(a)   (b)

A. DENSITY OF SULFUR Obtain several pieces of dry sulfur. Weigh a piece of weighing paper to the nearest 0.01 g. Place the sulfur on the paper and determine the mass to the nearest 0.01 g. By subtraction, find the mass of the sulfur.

Fill your 10-mL graduated cylinder with about 5 mL of water and read the volume to the nearest 0.1 mL. Carefully place the previously weighed sulfur lumps into the cylinder. Agitate slightly to be sure that no air bubbles are trapped. Read the new position of the meniscus to the nearest 0.1 mL. (Return the damp sulfur lumps to the proper container.) Calculate the volume of the sulfur sample and the density.

B. DENSITY OF UNKNOWN SOLID Obtain a sample of an unknown solid. Find the mass of this sample and the volume by water displacement as described above for sulfur. Calculate the density of the unknown solid.

4. DENSITY AS A FACTOR

On the report sheet give the setup of the calculation and the calculated answer for the following.

(a) The volume of a 34.0 g sample of the unknown liquid in part 2.

(b) The mass of a 135 cm^3 sample of the unknown solid in part 3.

5. THICKNESS OF METAL FOIL

To directly measure the thickness of a sample of aluminum foil a special instrument is needed. However, the thickness can be estimated indirectly by viewing a sheet of foil as a thin rectangular solid having a length, width and height. The height is the thickness. Since density relates mass to volume, the volume of a rectangular sample can be found from its mass. If we know the length, width and volume of a rectangular solid it is possible to calculate the height.

Obtain a sheet of aluminum foil having a regular shape. Smooth the sheet on your laboratory book and, using a meter stick, carefully measure the length of the edge of the sheet to the nearest 0.1 cm. If the sheet is not square, measure the length of adjacent edges. Use the balance to find the mass of the sheet. Do not crease or crumple the sheet. Find the area, A, of the sheet from the length of the edge or edges. The area is the product of the length and the width, or the square of the length of the edge if the sheet is a square.

$$A = l \times w \quad or \quad A = l^2$$

Calculate the approximate thickness of the foil as follows. Assuming that the foil is a rectangular solid, the volume is given by the area of the face times the height (thickness). This can be expressed as

$$V = At$$

where V is the volume, A is the area, and t is the thickness. If the density of aluminum is 2.70 g/cm^3, the volume of the foil sample can be determined from the mass and the density. Once the volume is known, it can be used along with the area to find the thickness of the foil.

$$t = V/A$$

Calculate the thickness of the aluminum foil in centimeters. Express the thickness in terms of millimeters.

6. THICKNESS OF A COIN

The thickness of a coin can be indirectly estimated if we assume that a coin is a cylindrical solid. For such a solid the volume is given by the product of the area of the circular base and the height or thickness. The area of the base can be found by measuring the diameter of the coin. The volume of the coin can be found using the mass of the coin and the density of the metal in the coin. Then the thickness of the coin can be calculated using the area of the circular base and the volume.

Measure the diameter of a nickel or quarter to 0.1 cm. Now determine the mass of the coin. Estimate the thickness of the coin as follows. Assuming that the coin is a cylinder, the volume is calculated by the formula

$$V = \pi(d/2)^2 t$$

where V is the volume, π is 3.14, d is the diameter, and t is the thickness (height). Nickels, dimes and quarters are composed of nickel and copper metals. If the densities of the metals in the coin are known, the volume of the coin is found from the mass. Both Ni and Cu have densities of 8.9 g/cm^3 so this density can be used in the calculations. Once the volume is known, it can be used along with the diameter to find the thickness of the coin. Calculate the thickness of your coin.

$$t = \frac{V}{\pi(d/2)^2}$$

7. AREA OF PAPER

Standard sheets of 8 1/2 by 11 inch copy paper are available in the lab. Observe the label on the package and obtain a fresh sheet. Your instructor will give you a small piece of this standard paper of unknown area. On your report sheet record the number written on this paper. Determine the total area of this small sheet of paper and report your answer in square centimeters. Hint: Since the paper has a fixed thickness you can use a whole sheet to find the mass per square inch or square centimeter of paper.

8. TABULATED DENSITIES

The *Handbook of Chemistry and Physics* contains many useful reference tables including tables of the densities of many materials and chemicals. Refer to a table of densities in the Handbook and record the names and densities of three different materials or chemicals.

3 Elements, Compounds and Mixtures

Objectives

To observe some elements and compounds, to isolate some elements from compounds, to form some compounds, to form a gaseous compound, and to analyze a mixture.

Terms to Know

Heterogenous Mixture - A mixture made of distinct parts or pieces of matter.

Pure Chemical - A homogeneous form of matter that has the same properties throughout.

Element - A pure chemical that cannot be separated into simpler chemicals.
A fundamental or elementary form of matter.

Compound - A pure chemical that can be separated into two or more simpler chemicals. A compound is a chemical combination of elements.

Filtration - The process of separating a solid from a liquid by passing the mixture through a porous medium such as a piece of filter paper.

Evaporating Dish - A bowl-shaped porcelain vessel used to contain chemicals so that they can be heated.

Watch Glass - A round and concave piece of glass which can be used to cover an evaporating dish.

Discussion

The objects and materials in our environment are forms of matter. Heterogeneous mixtures are made up of distinct parts. These mixtures can often be easily separated. Matter can be separated and classified into various categories. A pure chemical is a homogeneous form of matter that has the same properties throughout. There are two types of pure chemicals. Compounds are pure chemicals that can be separated into simpler chemicals called elements. Elements are pure chemicals that cannot be separated into simpler chemicals. Elements are elementary or fundamental forms of matter. Compounds are formed through chemical combinations of elements. Elements are isolated from compounds by the use of special chemical processes.

In this exercise you will observe the appearances and natures of some common elements and compounds. A mixture will be analyzed by separating the components and determining the percent of the water insoluble component. Two different elements will be isolated from compounds, and a compound will be made from its elements. A colorless, gaseous compound will be formed and its properties observed.

Laboratory Procedure

1. ANALYZING A MIXTURE

In this part you are going to analyze a mixture of salt and sand. When water is added to the mixture the salt will dissolve and the insoluble sand will not dissolve. The sand can be isolated by passing the water-sand mixture through a paper filter. Then the paper and sand can be dried. To express the composition of the mixture, in terms of the percent of sand, the mass of the sand and the mass of the original mixture are needed. The percent of sand is found by dividing the mass of the sand by the mass of the mixture and multiplying by 100.

Filtration involves passing the mixture through a membrane that allows the liquid to pass but retains the solid. Usually, the membranes used in filtration are made of specially prepared paper called filter paper. Such paper has small pores that allow the passage of the liquid only. To carry out the filtration process you need a funnel supported by a ring attached to a ring stand, as shown in Figure 3-1. The figure also shows how to fold the filter paper into a cone and how to filter a mixture. Set up the filtration system as shown in the figure. Obtain a piece of filter paper and fold it. Tear a tiny piece off of the inside corner as demonstrated by your instructor. Do not put the paper in the funnel at this time, since it needs to be weighed.

You will need an evaporating dish, a watch glass cover and a 50-mL beaker, a 100-mL beaker and a 250-mL beaker. Place the folded filter paper in the evaporating dish. Put the watch glass cover on the dish and weigh the dish, cover and paper to the nearest 0.01 g. Remove the folded filter paper and place it in the funnel. Wet the paper with some deionized water so that it fits snugly in the funnel.

Weigh a clean, dry 50-mL beaker to the nearest 0.01 g. Obtain a sample of the mixture to be analyzed from your instructor or from the bottle available in the laboratory. Place about 5 grams of the mixture into the 50-mL beaker and weigh the beaker and sample to the nearest 0.01 g. Pour about 30 mL of deionized water into the beaker and mix or stir to dissolve the salt in the mixture. Make sure that all of the salt has appeared to dissolve.

Place a 100-mL beaker below the funnel as shown in the figure. Be sure that the tip of the funnel is touching the side of the beaker. Carefully pour the water-sand slurry into the funnel to filter out the sand. Use a stream of water from your plastic wash bottle to wash any sand from the beaker into the filter paper in the funnel. Be sure that all of the sand is transferred to the filter paper cone in the funnel and be sure that no sand is dropped. Allow the water to flow through the funnel to the beaker. After most of the water has passed through the filter paper wash the paper and sand with a stream of deionized water from your wash bottle. Finally allow the water to drain through the paper.

To dry the paper and sand you need to set up an air bath as illustrated in Figure 1-1 on page 3. Carefully remove the wet filter paper from the funnel, fold it and place it in the evaporating dish. Put the watch glass cover on the dish and place it on the 250-mL beaker which will serve as an air bath to heat the dish. It is important to first adjust the height of the ring so that the hot part of the Bunsen flame will reach the bottom of the beaker. Light the Bunsen burner and heat the beaker with the full Bunsen flame. Allow the dish to heat until the paper is dry. The heating may take about 15 minutes. Continue heating until all of the water has evaporated, even the small drops which condense on the underside of the watch glass. While waiting for the heating to be completed go on with parts 2, 3, and 6 of this exercise.

After heating the dish, allow it to cool to room temperature. This will take about 10 minutes. When it is cool enough to touch you can move the dish to the top of the desk so that it will cool faster. The dish must be cooled to room temperature before you weigh it. When it is cool, weigh the covered dish to 0.01 g. As shown on the Report Sheet, use your data to calculate the percentage of sand in the mixture.

2. SOME COMMON ELEMENTS

Observe the samples of elements on display and fill in the table given on the Report Sheet.

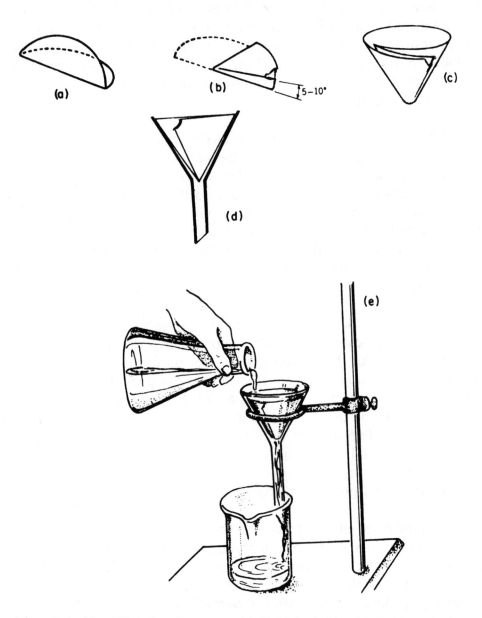

Fig. 3-1 The Filtration Process   (a) Fold in half. (b) Fold again but not exactly in quarters. Tear small piece from smaller fold. (c) Open to form a cone. (d) Place cone in funnel. Wet the paper with a small amount of water and seal the paper in the funnel. (e) Support the funnel and carefully pour the mixture to be filtered into the funnel. Never fill the paper full of liquid. Fill the filter about three-quarters full of liquid.

3. SOME COMMON COMPOUNDS

Observe the samples of compounds on display and fill in the table given on the Report Sheet.

PRECAUTION: In the next sections you will be using several chemicals and you will be heating chemicals. Always wear safety goggles or glasses when working with laboratory chemicals. Be careful of hot objects. Do not touch metals or glass ware that has just been heated. Sulfuric acid should not be spilled or splashed. If you splash acid on your clothing or skin, wash it off with large amounts of water and inform your instructor. See Parts 4 and 5 of Appendix 1 for a discussion of dispensing and heating of laboratory chemicals.

---

4. FORMING ELEMENTS FROM COMPOUNDS

(a) Silicon from Silica Sand

In this exercise the element silicon will be formed from silica sand. Silica sand contains a silicon-oxygen compound. When silica sand is heated with magnesium, the magnesium displaces the silicon in the compound forming elemental silicon.

Place several pieces of magnesium metal in the form of magnesium turnings or magnesium ribbon (about 0.2 g) in a clean, large test tube. Add enough silica sand to just cover the pieces of magnesium. Hold the tube with a test tube holder making sure that the holder is near the top of the tube. Hold the tube at an angle and place the bottom portion of the tube in the hot part of a Bunsen burner flame. Heat the tube in the flame for several minutes until you see the noticeable orange glow in the contents. After you see this glow in the contents, take the tube out of the flame and hold it while it cools for several minutes. Place the tube in a test tube rack or a small beaker until it cools to room temperature. Do not touch the hot tube. Go to part (b) of this exercise and return to this part after finishing part (b).

Obtain a clean piece of paper and pour the cooled contents of the tube onto the paper. Observe the contents and describe the difference in appearance of the contents before and after heating. Incidentally, the test tube used for this exercise will be permanently changed. Glass is also a compound of silicon and some silicon will be formed on the surface of the glass.

(b) Copper from a Copper-Oxygen Compound

In this exercise the element copper will be isolated from a copper-oxygen compound. When the compound is heated in the proper part of a flame and cooled quickly, copper metal will be formed. Copper metal can be recognized by its distinctive color.

Obtain a sample of copper-oxygen compound about the size of a pencil eraser. Place the sample on a piece of paper. Obtain a piece of copper metal in the form of a 10 cm length of 24 gauge wire. Place it on the piece of paper. Describe the appearance of the copper-oxygen compound and the elemental copper.

Fill a small beaker with cold water. Light your Bunsen burner and adjust the flame so that you can see a blue inner cone. Pick up a sample of the copper-oxygen compound about the size of a small pea using the tip of a stainless steel spatula. Carefully place the tip of the spatula in the Bunsen flame so that the sample is just above the blue inner cone. Heat the entire sample in this position for a few minutes. Move the spatula about to make sure that the entire sample glows orange and becomes very hot. Quickly immerse the tip of the spatula into the beaker of water. Remove the wet sample and place it on the piece of paper. Compare this sample to your sample of copper metal and describe the comparison. Save the piece of copper wire for part 5 of this experiment. Return to part (a) and finish your observations.

5. FORMING A COMPOUND FROM ELEMENTS

hood

Copper-Sulfur Compound

Copper and sulfur will be heated to form a distinctive copper-sulfur compound. Obtain a small piece of copper metal in the form of a 10 cm length of 24 guage wire. Place a sample of powdered sulfur about the size of a split pea on a piece of paper and transfer the sulfur into a test tube. Coil the copper wire and drop it into the test tube. Set up a Bunsen burner in the fume hood. Using your test tube holder, place the bottom of the test tube in the hot part of the Bunsen burner flame. Heat the bottom of the test tube with the hot part of the flame until the sulfur appears to be gone. Remove the tube from the flame, let it cool and observe it. After the tube has cooled remove the product. Describe the difference between the elements copper and sulfur and their compound which has formed.

6. A GASEOUS COMPOUND: CARBON DIOXIDE

Some compounds are gases and unless they have a color, it is not possible to see them. Such colorless gases can be detected by their chemical behavior. In this section, the gaseous compound carbon dioxide will be formed and detected. For this experiment you will need a 250-mL beaker, a clean drinking straw and a test tube and a cork or rubber stopper for the tube.

Fill the test tube about one fourth full of limewater solution. Stopper the tube and shake it. Record any observations. Pour a sample of solid sodium bicarbonate (baking soda) into the 250-mL beaker to cover the bottom of the beaker to a height of about 0.5 cm. Measure about 10 mL of dilute sulfuric acid into a graduated cylinder. This is a special solution of sulfuric acid and you find it in a bottle labeled 3 M sulfuric acid. CAUTION: Handle this acid solution with care. Pour the acid into the beaker containing the sodium bicarbonate. Note the vigorous bubbling as the carbon dioxide gas is formed. Since the gas is colorless, you will not be able to see it fill the beaker as it pushes the air aside. Remove the stopper from the test tube of limewater and tip the beaker over the top of the tube to pour some carbon dioxide into the tube. Do not allow any liquid in the beaker to enter the tube. Stopper the tube and shake it. Record your observations.

Remove the limewater solution from the test tube, rinse the tube with water and fill it with one fourth full of a fresh sample of limewater. Use a clean drinking straw to gently blow into the limewater in the tube. CAUTION: Be sure to use a clean straw and do not suck on the straw. Record your observations.

4 Empirical Formula

Objective

To carry out a chemical reaction that forms a chemical compound, and to determine the empirical formula of the compound.

Terms to Know

Empirical Formula - The simplest formula of a compound as deduced from the experimentally determined composition of the compound.

Molar Ratio - The ratio of the number of moles of one element to another element in a compound.

Crucible - A cup shaped porcelain vessel with a cover used to contain chemicals so that they can be heated.

Discussion

The formula of a compound shows which elements are in the compound and the relative number of combined atoms of each element. For example, the formula $CaCl_2$ indicates that the compound contains one combined atom of calcium for every two combined atoms of chlorine. From a molar point of view, the formula indicates that a mole of the compound contains two moles of combined chlorine atoms for each mole of combined calcium atoms. This fact can be expressed as a molar ratio

$$\frac{2 \text{ mole Cl}}{1 \text{ mole Ca}}$$

Such a molar ratio indicates the subscripts to be used in the formula (i.e., Ca_1Cl_2; the number 1 is omitted when writing the formula).

The simplest formula of a compound can be found by determining the amount in grams of each element present in a sample of a compound. This is done by reacting known amounts of elements with one another to form a compound or by separating the elements present in a sample of a compound and determining the amount of each individually. Once the amounts in grams of each element in a sample of a compound are known, the number of moles of each element can be calculated. The molar ratios can be found from the number of moles and used to deduce the empirical formula of the compound.

Consider an example of the experimental determination of the empirical formula of a compound. (Do the computations and fill in the blanks in the following discussion.) The example will concern the formation of a compound between gallium, Ga, and oxygen, made by heating a sample of gallium in air. By experiment it is found that a 1.29 g sample of gallium forms a product having a mass of 1.73 g. Assuming that the product is a gallium-oxygen compound, the mass of oxygen combined with gallium is found by subtracting the gallium mass from the product mass.

$$\text{Mass of combined oxygen} = 1.73 \text{ g} - 1.29 \text{ g} = \underline{0.44 g}$$

The number of combined moles of each element can be found from the masses.

$$\text{Moles of oxygen:} \quad \underline{0.44} \text{ g} \frac{1 \text{ mole O}}{16.00 \text{ g}} = \underline{2.8 \times 10^{-2} \text{ mol}}$$

$$\text{Moles of gallium:} \quad 1.29 \text{ g} \frac{1 \text{ mole Ga}}{69.72 \text{ g}} = \underline{1.85 \times 10^{-2} \text{ mol}}$$

Dividing the larger number of moles by the smaller will give the molar ratio of the two elements. (Use the number of moles from above to find the ratio.)

$$\frac{2.8 \times 10^{-2} \text{ moles O}}{1.85 \times 10^{-2} \text{ moles Ga}} = \frac{1.5 \text{ moles O}}{1 \text{ mole Ga}}$$

The formula using this ratio converted to a fraction is:

Ga $\underline{1}$ O $\underline{1.5}$

Since the ratio is not a whole number, each subscript should be multiplied by the number $\underline{2}$. This gives the final empirical formula of:

Ga $\underline{2}$ O $\underline{3}$

The empirical formula of a magnesium-oxygen compound will be determined in this experiment. The magnesium-oxygen compound is made by reacting magnesium metal with oxygen in the air. This will be accomplished by weighing a sample of magnesium and reacting it with oxygen in a closed crucible. The mass of the product is then determined, and the mass of oxygen that reacted with the magnesium can be found by subtracting the magnesium mass from the product mass. One problem is that at high temperatures some nitrogen will also react with the magnesium. This problem is overcome by reacting the magnesium-nitrogen compound with water and heating to convert to the magnesium-oxygen compound.

Laboratory Procedure

Remember to always wear safety goggles when working with chemicals. Wash, rinse and dry your crucible and cover. The crucible may have a dark stain in it from previous use but it can be used for this experiment. Be sure that the crucible does not have any small cracks in it. Place the crucible with the cover on a clay triangle supported on a ring stand as shown in Figure 4-1. Place your Bunsen burner under the ring so that the hot part of the flame reaches the bottom of the crucible. After heating, remove the burner and allow the crucible to cool to room temperature. Do not touch the hot crucible and do not put a hot crucible on the balance or your desk top. While the crucible is cooling, obtain about 50 cm of magnesium ribbon. (If necessary, clean any scale off the ribbon with steel wool or sand paper and wipe the ribbon clean with a paper towel or tissue.) Fold the magnesium sample into an accordion-like shape with about 1 to 2 cm length folds. It should be folded so that it will fit into the bottom of the crucible.

Weigh the cooled crucible and cover to the nearest 0.01 g. Record the mass on the report sheet. Push the folded magnesium sample into the bottom of the crucible and weigh the crucible, magnesium and cover. Record the mass. Cover the crucible and place it back on the clay triangle. Heat the crucible gently at first by moving the flame back and forth under the crucible for a few minutes. Then heat the crucible with the full flame for 15 minutes. The hot part of the flame must touch the bottom of the crucible so that it becomes so hot it turns orange in color. (DANGER: Do not heat the crucible with the cover off and do not remove the cover while heating.) After 15 minutes of heating, allow the crucible to cool for about 5 minutes and then, using your crucible tongs, adjust the cover so that there is a very small opening to the crucible. (See Fig. 4-2.) Heat the crucible using the full flame for 10 more minutes, remove the flame, and cool for a few minutes. Cautiously raise the cover with your crucible tongs and hold the lid just above the crucible. Heat the crucible with the full flame. If the contents of the crucible glow brightly and white smoke rises, place the cover on and heat for 5 minutes. Repeat the process of raising the cover, observing the product and replacing the cover until all of the magnesium has appeared to react. This will be apparent when the contents do not glow brightly or smoke when red hot. After you think the magnesium has reacted, heat with the full flame for a few minutes while holding the cover off. Replace the cover and allow to cool to room temperature.

After you are sure the crucible is cool, add 10 to 15 drops of deionized water to the crucible, cover and gently heat for 5 minutes, using a low flame. Water is added to convert any magnesium-nitrogen compound to magnesium-oxygen compound. Allow the crucible to cool to room temperature. The crucible will be cool enough to weigh if you can touch it and it does not feel warm. Finally, weigh the crucible, cover and product to the nearest 0.01 g and record the mass.

Use the data recorded on the report sheet to calculate the mass of magnesium and the mass of oxygen. Use these masses to determine the empirical formula of the magnesium-oxygen compound.

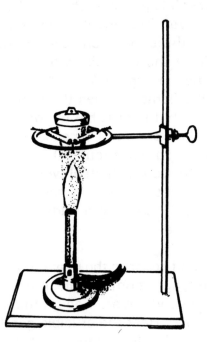

Fig. 5-1

Fig. 5-2

6 Graphing Data

Objective

To learn how to graph data and practice graphing methods

Terms to Know

X-axis - The horizonatal direction on a graph.

Y-axis - The vertical direction on a graph.

Slope of Straight Line - An expression of how the Y variable varies with the X variable.

Discussion

Sometimes we collect data in the laboratory that represents how one property depends upon another. For example, we may observe how the volume of a gas sample changes with the temperature. The observed properties are called variables and we measure various values of these variables by experiment. Normally in an experiment one variable is changed in a controlled fashion and corresponding values of the other variable are measured. For instance, we may change the temperature of a gas sample and measure the corresponding volumes. The variable for which the changes are controlled is called the independent variable. The other variable, that changes with the independent variable, is called the dependent variable. The temperature of the gas sample is the independent variable and the volume is the dependent variable.

To show how the variables change with respect to one another it is possible to graph or plot the data on graph paper. The experimental data consists of a set of independent variable values and the corresponding dependent variable values. A data point is any two related variable values. Thus, the data consists of a set of data points. The plotting of data on graph paper involves labeling the graph, locating each point of the graph and connecting the points with a smooth curve. A straight line is drawn through the points if the data appears to fit a straight line or if we know that the data should be represented by a straight line.

A piece of graph paper contains a square grid with a horizontal axis called the X-axis and a vertical axis called the Y-axis. Normally, the independent variable is plotted on the X-axis and the dependent variable on the Y-axis. When data is graphed in this way it is said that the dependent variable is plotted versus the independent variable. On a piece of graph paper the long or short side of the grid can be used for either axes as needed. That is, the long side can be used for the X-axis or the short side can be used to best fit the data. If it is not obvious which variable is the independent variable any side can be used as desired.

The typical steps used in graphing or plotting data are:

1. Note the range of the variables. The range of each variable extends from the lowest value to the highest value. To find the range subtract the smallest value from the largest value. For the approximate range round the calculated range up to some convenient whole number.

2. Note the total number of divisions on each axis of the graph paper. Some paper has large divisions which are divided into smaller divisions.

3. Depending on the ranges of the variables decide which axis on the paper is to be the X-axis and which is to be the Y-axis.

4. Using the ranges of the variables decide how the lines on the axes are to be numbered. You have to decide what value each division represents. To decide, divide the range of values by the total number of divisions on the axis. This gives the value per division. However, for ease of graphing your data it is best to round the value per division to 1/2, 1, 2, 5 or 10 units per division or any other convenient value. Mark each axis with an appropriate sequence of numbers. Not all division need be numbered. For instance, you may number every 5 divisions or every 10 divisions. Sometimes axes are numbered starting with zero but this is not required. If your data does not extend to zero or the lowest value is not close to zero then you may want to start numbering the axes with a number that is near the lowest value rather than zero.

5. Label the axes of the graph with appropriate titles and show the units of the variables being plotted as part of the label.

6. Plot the data points on the graph paper. Locate each value of the independent variable on the X-axis and, then, move along the Y-axis to find the location of the corresponding value of the dependent variable. You need to pay close attention to the value of each division on the graph so that the values are located correctly. Mark a dot using a fine pencil or pen at the location of each point on the graph paper. Draw a small circle around each point on the graph to emphasize its location.

7. Once all of the points have been placed on the graph connect the points with a continuous line that best fits the points. If the variables seem to fall along a straight line or if you know that the variables should correspond to a straight line, draw the best straight line through the points. By best straight line is meant a line that lies as close as possible to all of the points. Use a ruler to draw a straight line. Some of the points may be located above or below the line but try to make the distances between the points and the line as small as possible. If the points do not seem to fit a straight line draw a smooth curved line between the points. Do not draw a crooked or jagged line by connecting the points with straight lines.

To illustrate graphing, consider plotting some experimental data. Observation of the number of cricket chirps per minute at various temperatures produced the following results (From Croxton, *Applied General Statistics*, Prentice-Hall, 1940).

Temperature	45 °F	48 °F	50 °F	53 °F	56 °F	62 °F	64 °F	66 °F
Chirps/min	32	42	50	61	72	94	101	109

A plot of this data will show the relationship between chirps/min and temperature.

The first step in constructing a graph is to obtain a piece of graph paper. Graph paper comes in a variety of forms. This discussion will be based upon a piece of graph paper like that shown in Fig. 5-1. The first step is to look at the range of the data and the divisions on the graph paper to decide how to number the divisions to fit the data. The temperature data ranges from 45 to 66 °F to give a range of about 30. The chirps per minute data ranges from 32 to 109 to give a range of about 80.

The long side of the graph paper is divided into nine large division each of which is divided into ten parts for a total of 90 divisions. If we divide the chirps/min range by this number of divisions we get 80/90 = 0.89 chirps/min per division. This is close to 1 chirp/min per division so it is possible to use the small divisions to each represent 1 chirp/min and each large division will then represent 10 chirps/min. Starting with 20 chirps/min and numbering to 110 chirps/min the Y-axis can be numbered as shown in Fig. 5-1.

The temperature data ranges roughly 30 °F. On the graph paper the X-axis is divided into 6 and 1/2 large divisions and each of these is divided into 10 parts to give a total of 65 divisions. Dividing the range by the number of divisions gives 30/65 = 0.46 °F per division. This ratio can be rounded to 0.5 °F per division. Using 0.5 °F for each small division makes 20 small divisions or two large divisions represent 10 °F. Using this approach and numbering from 40 °F to 70 °F the X-axis can be numbered as shown in Fig. 5-1.

Once we number the lines and label the axes (See Fig. 5-1), the data can be plotted. Each set of two corresponding values of the data will be a point on the graph. The first point is 32 chirps per minute at 45 °F. To locate this point move along the X-axis to 45, then move straight up along the Y-axis to 32. Keep in mind that each division on the X-axis represents 0.5 °F. Make a dot at this point as shown in Fig. 5-2. Put a small circle around the point to emphasize its location. The next point is 42 chirps/min at 48 °F. Move along the X-axis to 48 and, then, move straight up along the Y-axis to 42. Make a dot at this point and circle it. The process is repeated for each point. Fig. 5-2 shows all of the data points plotted except for 101 chirps/min at 64 °F. For practice, you can plot this point on the graph.

After all of the points have been plotted, we normally try to connect the points with as smooth a curve as possible. Such a curve may be a straight line or a curved line which ever seems to fit the data. The graph of the data shown in Fig. 5-2 seems to fit a straight line. Fig. 5-3 shows the graph with a straight line drawn through the data points. It appears, over the temperature range of the data, that a linear or straight line relation exists between the temperature and the chirps per minute. It can be stated that the number of cricket chirps per minute is directly proportional to the Fahrenheit temperature.

As another example of graphing consider the following data which represents the way in which the density of a gas changes with the temperature.

Temperature (°C)	0	51	102	148	204	250	301	349	399	453	504
Density (g/L)		1.29	1.08	0.94	0.83	0.74	0.67	0.61	0.57	0.53	0.49	0.45

Considering the range of the data (about 0 to 600 and about 0 to 1.3), the divisions on the graph paper can be numbered as shown in Fig. 5-4. Once the axes are numbered and labeled, the points can be plotted and a smooth curve can be drawn through the points as shown in the figure. Note that the Y-axis was labeled using a range of 1.30 g/L divided by 90 divisions (1.30/90 = 0.014 g/L per division). For convenience this ratio is rounded to 0.02. This means that each division represents 0.02 g/L so every five divisions corresponds to 0.1 g/L. Sometimes, it is necessary to split up the divisions so that they represent some fraction rather than a whole number. On the X-axis there are 65 divisions so the range 600 °C divided by 65 divisions rounds off to 10 °C per division. In the plot given in Fig. 5-4 the point at 250 °C and 0.67 g/L was not plotted. For practice, you can plot this point on the graph. The plot of density versus temperature gives a curved line from which we can conclude that the density is inversely proportional to the temperature. We know this because as the temperature increases the density decreases.

If a plot of data corresponds to a straight line, the line represents how the variable on the Y axis varies with respect to the variable on the X axis. The slope of the straight line represents the ratio of the one variable to the other. One way to find the slope of a straight line on a graph is by use of the following formula:

$$m = \frac{Y_2 - Y_1}{X_2 - X_1}$$

where m is the slope and the Y and X values correspond to two points on the graph which are on the line but far apart. For example the slope of the straight line shown on Fig. 5-3 is

$$m = \frac{109 - 32}{66 - 45} = \frac{77}{21} = 3.7$$

Expressing the slope using the units of the variables gives the ratio of 3.7 chirps per min/°F. The slope expresses the way in which the chirps per minute changes with temperature. The change is 3.7 chirps per minute for every Fahrenheit degree change.

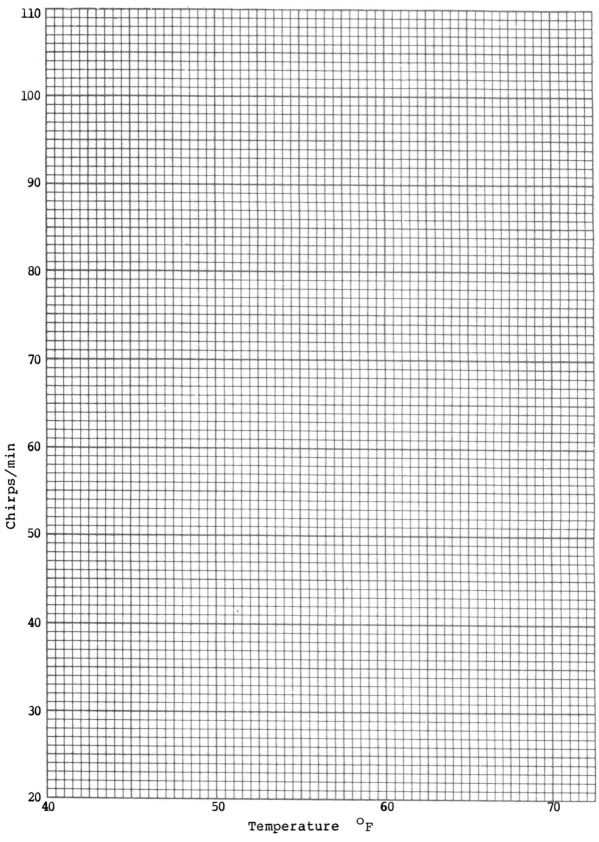

Fig. 6-1

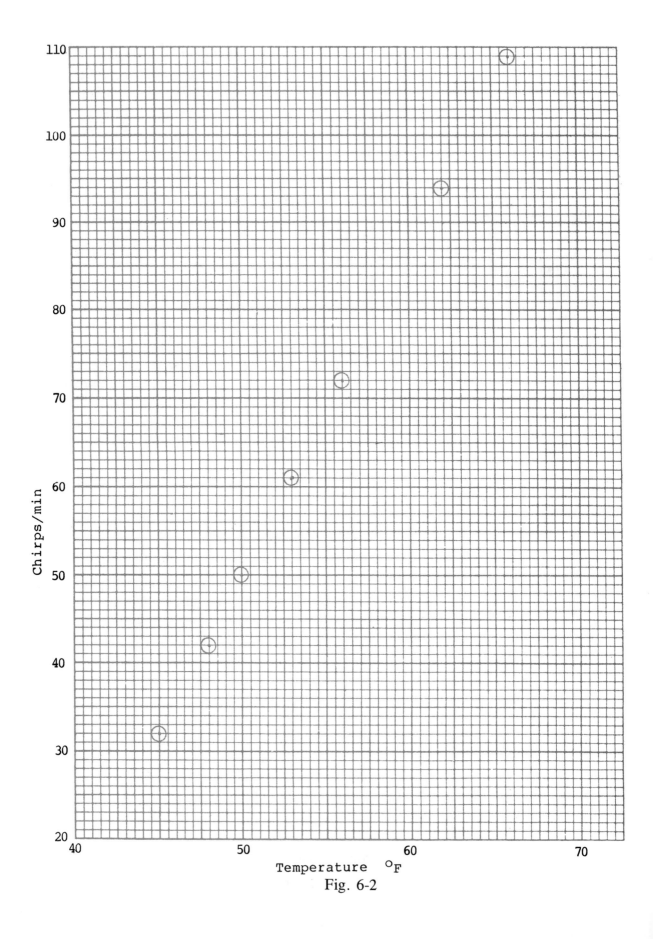

Fig. 6-2

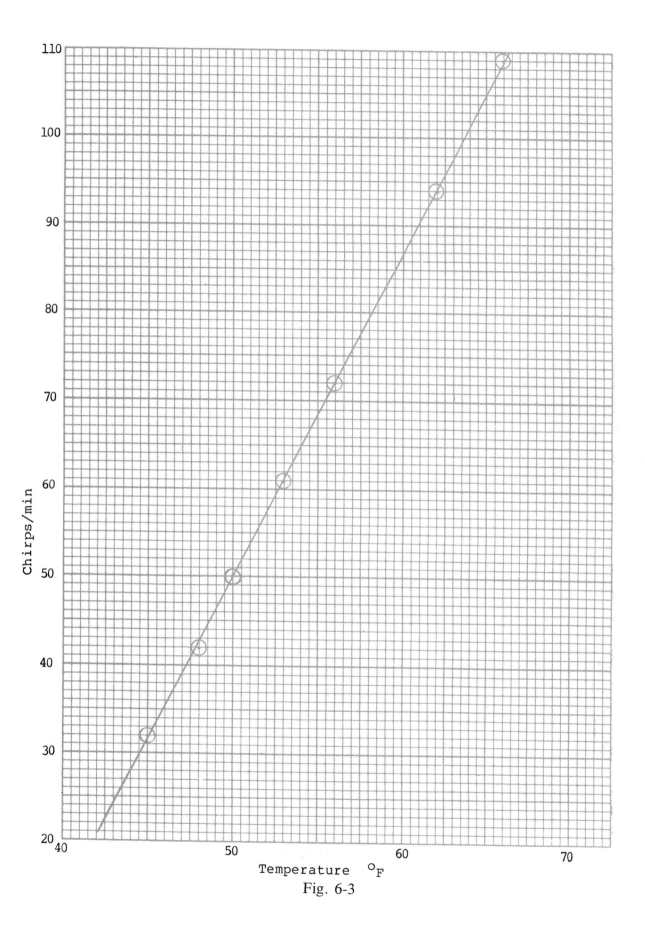

Fig. 6-3

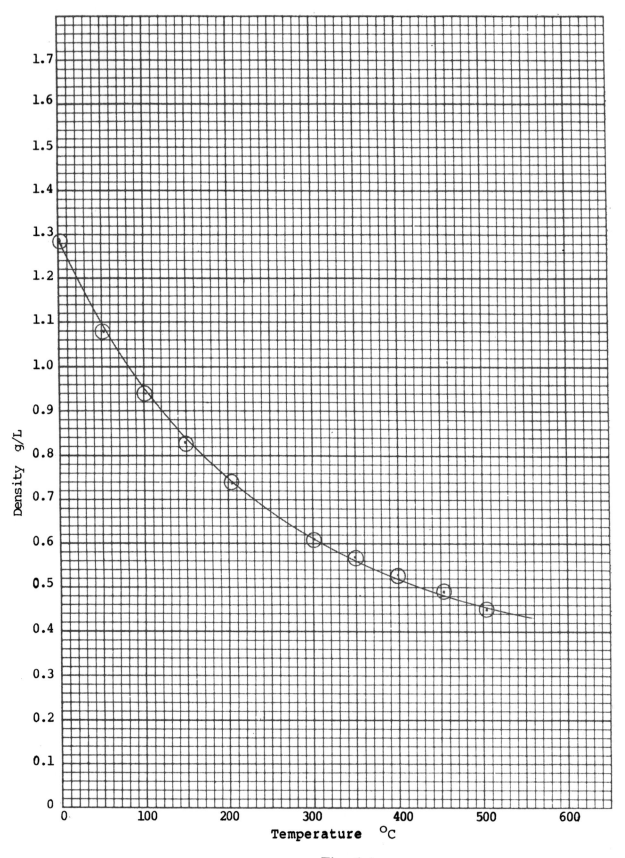

Fig. 6-4

Laboratory Procedure

In this exercise you will collect some experimental data and plot the data on graph paper. For the exercise you will need a laboratory balance, a 10-mL graduated cylinder and a 10-mL sample of a liquid. Obtain the liquid sample in a clean, dry 50-mL beaker. You want to measure the masses of various volumes of the liquid and plot the mass in grams versus the volume in milliliters. Collect the data as described and record your data in the space provided. The volumes should range up to but not exceeding 9 mL.

1. Weigh a clean, dry 10-mL graduated cylinder to the nearest 0.01 g.

2. Add about 1 mL of liquid, record the precise volume and weigh the cylinder plus the liquid to the nearest 0.01 g. Be sure that no liquid spills on the outside of the cylinder. If the cylinder gets wet on the outer surface just dry it off with a paper towel.

3. Add about 1 mL of the liquid to the cylinder along with the liquid that is already in the cylinder. Read the precise liquid level in the cylinder and record this volume. Weigh the cylinder plus the liquid to the nearest 0.01 g.

4. Repeat step 3 seven more times until you have measured and weighed a total of nine volumes of liquid. The last volume measured in the graduated cylinder should not exceed 9 mL.

5. To obtain the masses of the samples you will have to subtract the mass of the empty cylinder from each of your weighings.

Mass empty and dry cylinder _____

Mass of sample 1 + cylinder _____   Volume sample 1 _____
       - Mass cylinder _____

Mass of sample 1   _____

Mass of sample 2 + cylinder _____   Volume sample 2 _____
       - Mass cylinder _____

Mass of sample 2   _____

Mass of sample 3 + cylinder _____   Volume sample 3 _____
       - Mass cylinder _____

Mass of sample 3   _____

Mass of sample 4 + cylinder _____   Volume sample 4 _____
        - Mass cylinder _____

Mass of sample 4   _____

Mass of sample 5 + cylinder _____   Volume sample 5 _____
        - Mass cylinder _____

Mass of sample 5   _____

Mass of sample 6 + cylinder _____   Volume sample 6 _____
        - Mass cylinder _____

Mass of sample 6   _____

Mass of sample 7 + cylinder _____   Volume sample 7 _____
        - Mass cylinder _____

Mass of sample 7   _____

Mass of sample 8 + cylinder _____   Volume sample 8 _____
        - Mass cylinder _____

Mass of sample 8   _____

Mass of sample 9 + cylinder _____   Volume sample 9 _____
        - Mass cylinder _____
Mass of sample 9   _____

Summarize your data in the table given on the report sheet and use a piece of graph paper to plot the mass of the liquid samples (Y-axis) versus the volume (X-axis). Be sure to select the long and short axes of the graph paper so that you use as much graph paper as possible. This means that the graph paper can be turned to use the short side as the Y axis and the long side as the X axis if needed. Use a ruler or straight edge to draw the best straight line through the plotted points. The slope of the line gives the density of the liquid.

7 Some Chemical Reactions of Oxygen

Objective

The purposes of this experiment are to carry out some chemical reactions involving oxygen and other elements, to describe observations of the reactions, and to write balanced chemical equations for the reactions.

Terms to Know

Oxygen - An element that occurs as a colorless and odorless gas. It is a diatomic element represented in the elemental form as O_2.

Oxides - A general term used to refer to various compounds containing oxygen combined with other elements.

Chemical Reaction - A process in which one set of chemicals changes to another set of chemicals.

Chemical Equation - A symbolic representation of a chemical reaction which gives the formulas of the chemicals that react (reactants) and the formulas of the chemicals that form (products).

Combustion - A chemical reaction involving an organic compound and elemental oxygen that produces carbon dioxide and water. Often the release of heat and light accompany combustion.

Discussion

Chemical reactions are processes in which one set of chemicals react to form another set of chemicals. The initial set of chemicals are the reactants, and the chemicals produced are the products. The chemical reactions that an element undergoes with other elements and compounds are characteristic of that element and are the chemical properties of the element. It is not always possible to know that a chemical reaction has occurred when chemicals mix. However, an obvious change is a good suggestion that a reaction is taking place or has taken place. Consider an example. Sodium is a solid metallic element and chlorine is a green-yellow gas. When we mix these elements, the white crystalline solid sodium chloride (table salt) forms. The fact that a reaction occurred is apparent from the change in the appearance of the reactants compared to the different colored product. In addition, the formation of sodium chloride from sodium and chlorine is accompanied by the release of energy as light and heat. The release of energy also serves as an indication of the occurrence of a reaction. Other physical indications that a reaction may have occurred are a color change, the formation of a gas, or the formation of a distinct solid product.

As you carry out the reactions in this experiment, be alert and note any evidence of the occurrence of a chemical reaction. Describe each reaction by noting the nature of the reactants and products and any energy released as heat or light. A reaction is chemically described by writing a balanced equation. To write an equation, you need to know the formula of the product. The formulas of the products are given for the reactions in this experiment. As an example of the description of a reaction, consider the reaction of sodium and chlorine:

Solid metal + green gas produce white solid accompanied by heat and light

The balanced equation for the reaction is: $2Na + Cl_2 \longrightarrow 2NaCl$

Oxygen, which exists in the form of diatomic molecules, O_2, combines with most of the other elements to form binary compounds called oxides. A few examples of oxides are

Na_2O	Sodium oxide
MgO	Magnesium oxide
Al_2O_3	Aluminum oxide
CO_2	Carbon dioxide
N_2O_3	Dinitrogen trioxide
SO_2	Sulfur dioxide
Cl_2O_7	Dichlorine heptoxide

Since oxygen is present in the air, in some cases an oxide is formed by heating the element in a container open to the air. As an example of the reaction of oxygen with an element, consider the reaction of calcium. When solid calcium is strongly heated in air, it reacts with oxygen to give the white solid calcium oxide, CaO, or lime. The observations of this reaction are:

Solid metal + colorless gas produce white solid accompanied by the emission of light.

The balanced equation for the reaction is: $2Ca + O_2 \longrightarrow 2CaO$

Laboratory Procedure

Each of the following reactions involves some chemical reacting with the colorless gas oxygen. On the report sheet summarize your observations and give balanced equations. Formulas of any compounds involved in a reaction are listed in each part of the experiment. Remember that the element oxygen occurs as O_2 and we represent samples of most other elements by their symbols.

1. REACTIONS WITH OXYGEN IN THE AIR

A. ALUMINUM Aluminum metal readily combines with oxygen to form a protective oxide coating that prevents further reaction with oxygen. In fact, a property of discarded aluminum is that it forms aluminum oxide very slowly over a period of many years. However, it is possible to combine aluminum and oxygen more quickly in the presence of a catalyst.

Obtain a 1 cm length of aluminum wire. Vigorously rub one end of the wire with steel wool or sand paper to polish it. A small beaker of mercury(II) nitrate solution is available in the laboratory. Dip the polished end of the wire into the solution of mercury(II) nitrate catalyst. Place the wire on the edge of your laboratory book with the dipped end protruding over the edge.

If wire is not available, obtain a small piece of aluminum foil and clean the surface with a piece of steel wool. Place a drop of mercury(II) nitrate solution on the cleaned surface of the foil to serve as a catalyst. Allow the aluminum to sit undisturbed for 5 to 10 minutes. (Continue with the other parts of the experiment and return to this part in 5 or 10 minutes.)

Record any observations and write a balanced equation for any reaction that occurred between aluminum and oxygen. The product is aluminum oxide, Al_2O_3. The mercury(II) nitrate is a catalyst and is not included in the equation as a reactant or a product.

B. MAGNESIUM Obtain a 2 cm length of magnesium ribbon. Hold the ribbon with tweezers or crucible tongs and place it in the Bunsen burner flame. (DANGER: To protect your eyes, do not look directly into the flame when this reaction occurs but turn your head and look out of the corners of your eyes.) Record any observations and write a balanced equation for any reaction that occurred between magnesium and oxygen. The product is magnesium oxide, MgO.

C. COPPER Obtain a small length of copper sheet and a small beaker of cold water. Hold the sheet with crucible tongs or tweezers and place it in the Bunsen burner flame. Move the sheet around in the flame and observe its appearance. Remove it from the flame and quickly dip it into the beaker of water. Observe the appearance of the sheet. Once again hold the copper in the Bunsen flame until it is very hot. Remove it from the flame and let it cool in the air. Record the appearance of the sheet. Two reactions took place. The compound copper(I) oxide, Cu_2O, has a reddish color and copper(II) oxide, CuO, has a black-grey color. Write a balanced equation for each reaction between copper and oxygen.

D. HYDROGEN Place a few pieces of mossy zinc in a small test tube. Pour about 1 mL of dilute hydrochloric acid (6 M HCl) into the tube. As the reaction occurs, hydrogen gas will bubble off. Allow the gas to bubble off for a moment. Now invert a large test tube over the top of this tube and fill it with the rising hydrogen gas. Allow the tube to fill with hydrogen for a minute or two. Stopper this hydrogen filled tube with a cork or rubber stopper and then rinse out the small test tube to stop the hydrogen generation. Do not throw the unreacted zinc in the sink.

Obtain a wooden splint and ignite it in the Bunsen flame. Remove the stopper from the hydrogen tube and touch the flaming splint to the mouth of the tube. (CAUTION: This is a violent reaction, so do not get too close to the tube and do not put your face close to the tube.) If you do not observe a reaction, collect another sample and try again. Record any observations and write a balanced equation for any reaction between hydrogen and oxygen. Hydrogen is a diatomic element represented as H_2 and the product of the reaction is water, H_2O.

2. COMBUSTION

In this exercise the term combustion refers to the reaction of a carbon-containing compound (organic compound) with oxygen to form carbon dioxide, CO_2, and water, H_2O. Equations for such combustion reactions always include the organic compound and O_2 as reactants, and carbon dioxide and water as products.

$$\text{compound} + O_2 \longrightarrow CO_2 + H_2O$$

A. NATURAL GAS Natural gas is a mixture of gases called hydrocarbons; but the main component is methane, CH_4. When you light your Bunsen burner, the combustion of methane occurs, producing the expected combustion products accompanied by heat and light. You can detect the production of water during the combustion of methane by inverting a dry test tube over the edge of the Bunsen burner flame. Hold the test tube at a 45 degree angle to the edge of the flame, using a test tube holder. The water will condense on the sides of the tube. Write a balanced equation for the combustion of methane and describe the reaction.

B. CELLULOSE Cotton fiber consists mainly of the compound cellulose. Place a loose wad of cotton about the size of a marble on a ceramic pad or wire gauze. Ignite the cotton with a Bunsen flame. Cellulose is a complex chemical compound that we can represent by the empirical formula, $C_6H_{10}O_5$. Write a balanced equation for the combustion of cellulose and describe the reaction. The small amount of ash comes from impurities in the cotton and some unburned carbon. Do not consider the ash as a product of the main reaction.

C. ISOPROPYL ALCOHOL *Ethanol* Work in a fume hood containing a Bunsen burner. You need some tweezers. Obtain a packaged isopropyl alcohol swab. (If swabs are not available use a 5 cm square of paper towel that is totally wet with isopropyl alcohol.) Open the package and hold the swab with the tweezers. Touch the swab to the flame to ignite it and remove it from the flame. Allow the alcohol to burn until the flaming stops. Record your observations and write a balanced equation for the combustion of isopropyl alcohol, C_3H_8O. *C_2H_5OH*

3. REACTIONS WITH PURE OXYGEN

A. PREPARATION OF PURE OXYGEN Samples of pure oxygen can be prepared by decomposing hydrogen peroxide, H_2O_2. A water solution of hydrogen peroxide is a source of this reactant. In this reaction, in which yeast is a catalyst, hydrogen peroxide decomposes to form water and oxygen gas. Give the balanced equation for this reaction.

Set up the gas production apparatus shown in Fig. 6-1. You will need a bent and a circular piece of glass tubing. See Part 6 of Appendix 1 for a discussion of glassworking and inserting a tube into a rubber stopper. Completely fill two large test tubes with water and stopper them. Invert the tubes into a 250-mL beaker half filled with water. Remove the stoppers from the test tubes. Pour about 20 mL of a fresh 3% hydrogen peroxide solution in the test tube in the gas production apparatus. Then add a sample of solid yeast about the size of a marble and seal the tube with a stopper attached to rubber tubing.

Collect two test tubes of oxygen gas by displacing the water in the inverted test tubes. Allow the oxygen gas to bubble into the tubes until all of the water is displaced. As you fill a tube with gas, remove it and stopper it tightly. If the bubbling of oxygen slows too much remove the test tube from the apparatus, dump out the liquid and refill the tube with more hydrogen peroxide solution and add more catalyst. After filling two tubes, rinse out the gas generating tube to stop the reaction.

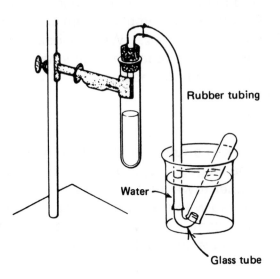

Fig. 7-1 Gas generation apparatus

B. CARBON   Obtain a small stick of charcoal (assume that it is carbon, C) and hold it with your tweezers. Briefly heat the charcoal in the Bunsen burner flame to ignite it. Quickly remove the stopper from a tube of oxygen and immerse the tweezers into the oxygen. Do not drop the burning charcoal into the tube. Record your observations and write a balanced equation for the reaction between carbon and oxygen to give carbon dioxide, CO_2. The white ash that may remain is a result of impurities in the charcoal and is not a product of the main reaction.

C. IRON   Obtain a small piece of steel wool and twist it into a long wad about 2 or 3 cm long. Holding the wad with your tweezers, heat it in the Bunsen burner flame for a few seconds to get it hot. Quickly plunge it into a tube of oxygen. Hold on to the tweezers and do not drop the burning steel wool into the tube. Record your observations and write a balanced equation for the reaction between iron and oxygen. The products may appear as a smoke or fine solid. Iron can form three oxides: FeO, F_2O_3 and Fe_3O_4. Write a separate equation for the formation of each of these oxides.

8 Hydrates: Water in a Complex Compound

Objective

The purposes of this experiment are to study the behavior of some hydrates and to determine the percent-by-mass of water in a hydrate.

Terms to Know

Percent-by-Mass - The number of grams of a part in 100 parts of the whole. The percent-by-mass of a component is determined by dividing the mass of a component by the mass of the whole and multiplying by 100.

Hydrate - A complex chemical compound made up of one compound combined in definite proportions with water so that there is a specific number of moles of water per mole of the compound.

Crucible - A cup shaped porcelain vessel with a cover used to contain chemicals so that they can be heated.

Discussion

Water is one of the most common chemicals on earth. It is found underground, in lakes, in rivers, and in the oceans. It is even found in the atmosphere as clouds and water vapor. It is put to many uses and it is fundamental to life processes. Water is sometimes found as a component of complex compounds called hydrates. Hydrates are interesting kinds of complex compounds formed by the inclusion of water into the crystals of a solid. Some minerals are hydrates. Examples are gypsum, borax and

Epsom salts. Hydrates are complex compounds because they are compounds composed of two other compounds, one of which is water. In fact, the word hydrate means "combined with water." A hydrate is a complex compound containing a specific number of moles of water per mole of the compound. Formulas for hydrates are written by first giving the formula of the compound other than water. This is followed by a dot and the formula of water preceded by a number showing the number of moles of water. For example, a hydrate of calcium chloride contains two moles of water per mole of compound. Thus, the formula is $CaCl_2 \cdot 2H_2O$. Some other examples of hydrates are:

$CuSO_4 \cdot 5H_2O$ Copper(II) sulfate pentahydrate
$CaCl_2 \cdot 2H_2O$ Calcium chloride dihydrate
$MgSO_4 \cdot 7H_2O$ Magnesium sulfate heptahydrate
$Na_2CO_3 \cdot 10H_2O$ Sodium carbonate decahydrate

The water present in hydrates is called water of hydration. It is sometimes possible to remove the water of hydration (dehydrate) a hydrate by gently heating it. Heating a hydrate produces water and the anhydrous (without water) solid compound. For example

$$CaCl_2 \cdot 2H_2O \xrightarrow{\text{heat}} CaCl_2 + 2H_2O$$
anhydrous
calcium chloride

The percent-by-mass water in some hydrates can be determined by dehydrating a known amount of the hydrate and then determining the mass of the resulting anhydrous solid. The mass of water is found by subtracting the mass of anhydrous solid from the mass of the original sample of hydrate:

mass H_2O = mass hydrate − mass anhydrous solid

The percent of water is then found by dividing the mass of water by the mass of the original sample of hydrate and multiplying by 100.

$$\% \ H_2O = \frac{\text{mass } H_2O}{\text{mass hydrate}} \times 100$$

Some anhydrous chemicals can combine with the water vapor in the atmosphere to form hydrates. When a chemical absorbs water from the atmosphere, it is said to be hygroscopic. Sometimes hygroscopic chemicals are used as drying agents or desiccants. Some chemicals are so hygroscopic that they absorb enough water from the atmosphere so that they dissolve and form a solution. Such chemicals are said to be deliquescent. In fact, deliquescent chemicals have to be stored in tightly sealed containers or they will dissolve in the water they absorb from the atmosphere. A few hydrates actually spontaneously lose water of hydration when open to the atmosphere. The spontaneous loss of water is called efflorescence and such hydrates are termed efflorescent chemicals.

Laboratory Procedure

1. DELIQUESCENCE AND EFFLORESCENCE

Place a few crystals of sodium sulfate decahydrate, $Na_2SO_4 \cdot 10H_2O$, on a sheet of paper or a watch glass. Place a few crystals of calcium chloride, $CaCl_2$, on the same sheet but as far from the other crystals as possible. Allow the crystals to remain on the sheet for about one hour and then record any changes you observe. (Continue with the rest of the experiment and return to this part later.) Which of the two chemicals is deliquescent? Which of the two chemicals is efflorescent?

2. EFFECT OF HEATING A HYDRATE

Place a few crystals of copper(II) sulfate pentahydrate in a dry test tube. Use a test tube holder to hold the test tube almost horizontally and gently heat the crystals by passing the tube through a Bunsen flame. Note any changes. When no further changes occur, allow the tube to cool. After cooling, add a few drops of water. Note any changes. Feel the tube and note any temperature change. Describe your observations.

3. THE PERCENT-BY-MASS WATER IN A HYDRATE

Clean a crucible with a cover and place the covered crucible in a clay triangle supported on a ring stand. (See Fig. 8-1.) Carefully heat the crucible and cover for a few minutes to make sure they are dry. Allow to cool to room temperature and determine the mass of the covered crucible to the nearest 0.01 g. Obtain a sample of hydrate. Place about 3 g of the hydrate in the crucible, cover, and weigh to the nearest 0.01 g.

Put the crucible back on the clay triangle and drive off the water by gently heating the crucible for about 15 minutes. Use a medium flame to heat the crucible and do not allow the crucible to become red hot. This can be done by making sure that the tip of the flame does not touch the bottom of the crucible, as shown in Fig. 8-1. After heating, cool to room temperature and determine the mass to the nearest 0.01 g. Gently heat the crucible for another 5 minutes, cool and weigh. If the first and second weighings do not agree within 0.03 g, reheat for 5 minutes and weigh again. Repeat until subsequent weighings agree.

The mass of water which was contained in the sample can be found by subtracting the final mass of the heated crucible from the mass of the crucible with the hydrate. The mass of the hydrate sample can be found by subtracting the mass of the empty crucible from the mass of the crucible with hydrate. The percent of water by mass in the hydrate can be found by dividing the mass of water by the mass of the hydrate sample and multiplying by 100.

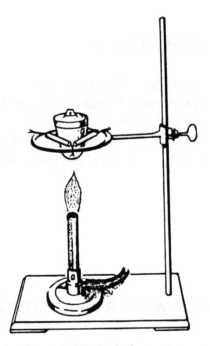

Fig. 8-1

4. THE FORMULA OF THE HYDRATE (OPTIONAL)

Show your percent of water results to your instructor and he or she will give you the formula of the parent compound contained in the hydrate. From the formula you can find the number of grams per mole of the parent compound. To determine the formula of the hydrate, calculate the percent of the parent compound in the hydrate by subtracting the percent of water from 100.

_____

Based on this percent, how many grams of parent compound would be contained in 100 g of hydrate?

_____

Use the molar mass of the parent compound to find the number of moles of this compound corresponding to the mass above.

_____

Using the percent water, how many grams of water would be contained in 100 g of the hydrate?

_____

Use the molar mass of water to find the number of moles of water in the sample.

_____

The formula of the hydrate is found by calculating the number of moles of water per one mole of the parent compound. This is done by dividing the number of moles of water by the number of moles of parent compound.

_____

The formula of the hydrate is

_____ _____H_2O

8 REPORT SHEET   EXPERIMENT 8

Name _____

Lab Section _____   Due Date _____

1. Deliquescence and Efflorescence

_____ efflorescent   _____ deliquescent

2. Effect of Heating a Hydrate

3. The Percent Mass Water in a Hydrate

Mass crucible and hydrate   _____

Mass crucible   _____

Mass heated crucible

_____   _____   _____   _____

Mass water   _____

Mass hydrate   _____

Percent by mass water   _____ x 100 = _____%

4. Formula of Hydrate (Optional)

Percent water   _____

Percent parent compound   _____

Grams parent compound   _____

Molar mass parent compound   _____

Moles parent compound

$$\underline{\hspace{3cm}} \quad \underline{\hspace{3cm}} \quad = \quad \underline{\hspace{3cm}}$$

Grams water   _____

Moles water

$$\underline{\hspace{3cm}} \quad \underline{\hspace{3cm}} \quad = \quad \underline{\hspace{3cm}}$$

Moles of water per mole of parent

$$\underline{\hspace{3cm}} \quad = \quad \underline{\hspace{3cm}}$$

Formula of hydrate   _____ · _____H_2O

8 QUESTIONS   EXPERIMENT 8

Name _KC Anderson_

Lab Section _____   Due Date _____

1. Explain what effect each of the following factors has on the determination of the percent of water in a hydrate. Indicate whether the factor will give a high or low percent and explain why.

    (a) The hydrate was heated too strongly and some of the product decomposed to give a gaseous product. _water percentage will be higher due to the loss of some of the substance being lossed when the gaseous product was produced._

    (b) The hydrate was not heated enough and some water remained after the final heating. _water percentage will be lower due to the additional mass recorded following drying._

2. Plaster of Paris used in plaster casting is made by gently heating gypsum, $CaSO_4 \cdot 2H_2O$. Actually Plaster of Paris is a also a hydrate, but one that contains less water of hydration than gypsum. When water is added to powdered Plaster of Paris, it reforms solid gypsum. If Plaster of Paris is a hydrate of calcium sulfate containing 6.18% water, what is the formula of the Plaster of Paris hydrate?

    _6.18% H_2O_

3. Using your formula for Plaster of Paris obtained in question 2 and the formula for gypsum shown in question 2, write a balanced equation that shows the reaction between Plaster of Paris and water to form gypsum.

4. A desiccant is a chemical that is used to keep a closed container or a confined space free from excess water vapor. Typical desiccants are silica gel and calcium chloride. Explain how desiccants work in terms of hydrate formation.

9 Stoichiometry

Objective

The purpose of this exercise is to experimentally confirm some principles of stoichiometry.

Terms to Know

Stoichiometry - Calculations that involve mass relations between reactants and/or products in a chemical reaction.

Molar Ratio - The ratio of the number of moles of one species involved in a reaction to the number of moles of another species in the reaction as revealed by the coefficients in the equation for the reaction.

Molar Mass - The number of grams per mole of a species as obtained from its formula.

Evaporating Dish - A bowl shaped porcelain vessel used to contain chemicals so that they can be heated.

Watch Glass - A round and concave piece of glass which can be used to cover an evaporating dish.

Discussion

Chemical stoichiometry deals with mass relations in chemical reactions. Stoichiometry is based upon the fact that mass is conserved in a chemical reaction. This means that the total mass of the products in a reaction equals the total mass of the reactants used in the reaction. A balanced equation for a reaction reflects the conservation of mass. Stoichiometry relates various reactants and products through the number of moles of each species involved in a reaction. The coefficients in a balanced equation reveal the relative number of moles of each species involved in a reaction. The coefficients give the molar ratios between any two species.

In this experiment you will react a sample of sodium hydrogen carbonate with sulfuric acid. The products of the reaction are sodium sulfate, water and carbon dioxide. The sodium sulfate will be isolated by heating to drive off the carbon dioxide and water. An equation for the reaction is

$$2NaHCO_3 + H_2SO_4 \longrightarrow Na_2SO_4 + 2H_2O + 2CO_2$$

Any two species in the reaction can be related by molar ratios revealed by the coefficients. For example, sulfuric acid and sodium hydrogen carbonate are related by the ratios:

$$\frac{2 \text{ mol } NaHCO_3}{1 \text{ mol } H_2SO_4} \quad \text{and} \quad \frac{1 \text{ mol } H_2SO_4}{2 \text{ mol } NaHCO_3}$$

Sulfuric acid and carbon dioxide are related by the ratios:

$$\frac{1 \text{ mol } H_2SO_4}{2 \text{ mol } CO_2} \quad \text{and} \quad \frac{2 \text{ mol } CO_2}{1 \text{ mol } H_2SO_4}$$

Molar ratios are used as factors to determine the number of moles of one species related to a given number of moles of another species. For instance, to determine the number of moles of CO_2 formed when 0.378 moles of H_2SO_4 react, we multiply the moles of H_2SO_4 by the appropriate factor:

$$0.378 \text{ mol } H_2SO_4 \; \frac{2 \text{ mol } CO_2}{1 \text{ mol } H_2SO_4} \quad = \quad 0.756 \text{ mol } CO_2$$

The number of moles of any substance is related to its mass using the molar mass as a factor. Using molar masses and molar ratios, a given mass of any species in a reaction can be related to the mass of any other species. For example, suppose we want to determine the number of grams of CO_2 formed when 0.690 g of H_2SO_4 react. Since the equation reveals the relation between the two species on a molar basis, we must work with moles. First, the molar mass of H_2SO_4 is used to find the number of moles of H_2SO_4 in the given mass. The molar ratio is then used to find the corresponding number of moles of CO_2. Finally, the molar mass of CO_2 is used to find the mass of CO_2 formed. The molar masses that are required in a calculation have to be deduced from the formulas. Consider the steps involved in the calculations. First, the number of moles of H_2SO_4 is found from the mass.

$$0.690 \text{ g} \; \frac{1 \text{ mol } H_2SO_4}{98.07 \text{ g}}$$

This product is multiplied by the molar ratio relating CO_2 to H_2SO_4 to find the moles of CO_2.

$$0.690 \text{ g} \; \frac{1 \text{ mol } H_2SO_4}{98.07 \text{ g}} \; \frac{2 \text{ mol } CO_2}{1 \text{ mol } H_2SO_4}$$

Finally, the molar mass of CO_2 is used to find the grams of CO_2.

$$0.690 \text{ g} \; \frac{1 \text{ mol } H_2SO_4}{98.07 \text{ g}} \; \frac{2 \text{ mol } CO_2}{1 \text{ mol } H_2SO_4} \; \frac{44.01 \text{ g}}{1 \text{ mol } CO_2} \; = \; 0.619 \text{ g}$$

In this experiment you will carry out the reaction of sodium hydrogen carbonate and sulfuric acid and collect the sodium sulfate formed. The mass of the sodium sulfate collected will be used in several stoichiometric calculations.

Laboratory Procedure

For the experiment you will need a 100-mL beaker, a 250-mL beaker, an evaporating dish and a watch glass cover. Obtain a test tube of $NaHCO_3$ and pour the contents into the 100-mL beaker. Record the number of the sample. If any of the sodium hydrogen carbonate remains in the tube, rinse it into the beaker with a small amount of deionized water. Measure 7 mL of 1 M sulfuric acid in a graduated cylinder. Slowly pour the sulfuric acid into the beaker of sodium hydrogen carbonate. Be careful not to let the solution bubble out of the beaker as you add the acid. After you have added all of the acid and the bubbling has stopped, rinse the sides of the beaker with a small amount of deionized water using your wash bottle. Be sure to use a small amount of water.

Add two or three drops of methyl orange indicator to the beaker. Obtain a dropper bottle of sulfuric acid and add the acid to the beaker drop by drop until the color changes from yellow to orange. It is very important not to add too much acid at this point. If you think that you added too much acid or if the solution turns red rather than orange you will need to start the experiment over.

Place two boiling chips in an evaporating dish, put the watch glass cover on the dish and weigh this combination to 0.01 g. Pour the solution from the 100-mL beaker into the evaporating dish and cover with the watch glass. Place the covered dish on top of a 250-mL beaker supported by a ring stand as shown in Figure 9-1. The beaker will serve as an air bath to heat the dish. Adjust the height of the ring so that the hot part of the Bunsen flame will reach the bottom of the beaker. Light the Bunsen burner and heat the beaker with the full Bunsen flame. Allow the solution in the dish to gently boil until all of the water has boiled off. The heating may take some time. Continue heating until all of the water has evaporated, even the small drops which condense on the underside of the watch glass. IMPORTANT NOTE: If white fumes or smoke evolve from the dish near the end of the heating, stop the experiment and consult your instructor. After heating the dish, allow it to cool to room temperature. When it is cool enough to touch, you can move the dish to the top of the desk so that it will cool faster. The dish must be cooled to room temperature before you weigh it. When it is cool, weigh the covered dish to 0.01 g.

Use your experimental data to determine the mass of the sodium sulfate, Na_2SO_4, that was deposited in the dish. Using the mass of sodium sulfate and stoichiometric factors, calculate the number of grams of sodium hydrogen carbonate in your original sample, the number of grams of sulfuric acid used in the reaction, and the number of moles of carbon dioxide produced in the reaction.

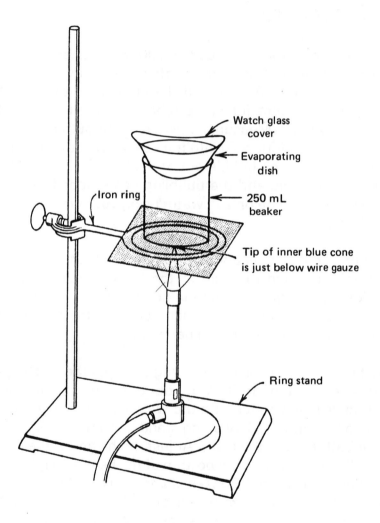

Watch glass cover

Evaporating dish

Iron ring

250 mL beaker

Tip of inner blue cone is just below wire gauze

Ring stand

Fig. 9-1 Apparatus for Air Bath

11 Analysis of Vinegar

Objective

The purpose of this experiment is to carry out a chemical analysis of the acetic acid content of vinegar and express the percent by mass acetic acid in vinegar.

Terms to Know

Vinegar - A water solution of acetic acid used as a condiment.

Acetic Acid - A common chemical which has the formula of $HC_2H_3O_2$ and belongs to a group of chemicals called acids. The active ingredient in vinegar.

Sodium Hydroxide - A chemical compound which has the formula of $NaOH$ and belongs to a group of compounds called bases.

Percent-by-Mass - The number of grams of a part in 100 grams of the whole. The percent-by-mass of a component is determined by dividing the mass of a component by the mass of the whole and multiplying by 100.

Discussion

Vinegar has been used for centuries as a sour food flavoring agent. Literally, vinegar means "sour wine" which suggests how it was first discovered. Vinegar is a water solution of acetic acid, $HC_2H_3O_2$. In this experiment you are going to analyze a sample of vinegar to find the percent-by-mass of acetic acid in the vinegar. To find the amount of acetic acid in a sample of vinegar, a solution of sodium hydroxide, NaOH, is added and the following reaction occurs:

$$HC_2H_3O_2(aq) + NaOH(aq) \longrightarrow NaC_2H_3O_2(aq) + H_2O$$

The solution of NaOH contains a known number of moles of dissolved NaOH per gram of solution. If we determine the number of grams of NaOH solution needed to react with the acetic acid in the vinegar, we can determine the number of moles of $HC_2H_3O_2$. This can be done by using the fact that 1 mole of NaOH is needed to react with 1 mole of $HC_2H_3O_2$. Once we know the number of moles of acetic acid, we can find the mass of the acid using its molar mass. To illustrate, suppose we analyze a vinegar solution using a NaOH solution which contains 0.00128 moles NaOH/g of solution. The experiment reveals that it requires 12.32 g of a NaOH solution to react with the acetic acid in a vinegar sample. The number of moles of NaOH involved in the reaction is found by multiplying the mass of the solution used by the number of moles of NaOH per gram of solution:

$$12.32 \text{ g } \frac{0.00128 \text{ mole NaOH}}{1 \text{ g}}$$

Next, the moles of acetic acid in the vinegar can be calculated by using the fact that 1 mole of $HC_2H_3O_2$ reacts with 1 mole of NaOH.

$$12.32 \text{ g } \frac{0.00128 \text{ mole NaOH}}{1 \text{ g}} \frac{1 \text{ mole } HC_2H_3O_2}{1 \text{ mole NaOH}}$$

Once we know the number of moles of acetic acid we can calculate the mass of the acid by multiplying by the molar mass:

$$12.32 \text{ g } \frac{0.00128 \text{ mole NaOH}}{1 \text{ g}} \frac{1 \text{ mole } HC_2H_3O_2}{1 \text{ mole NaOH}} \frac{60.05 \text{ g}}{1 \text{ mole } HC_2H_3O_2} = 0.947 \text{ g}$$

In the experiment, NaOH and vinegar solution are mixed. To determine when sufficient NaOH has been added, we place an indicator, named bromthymol blue, in the solution. Bromthymol blue turns blue in the presence of NaOH and is yellow in vinegar. If bromthymol blue is in the solution mixture and the NaOH solution is slowly added until the mixture turns blue (or green), this indicates that the proper amount of NaOH has been used.

Laboratory Procedure

For this experiment you need a 50-mL Erlenmeyer flask, a 50-mL beaker and a clean eye dropper. Bottles of sodium hydroxide solution and vinegar are available in the laboratory. **CAUTION: sodium hydroxide solutions are caustic and dangerous. Wear safety goggles.** Add 1 or 2 drops of bromthymol blue indicator to the flask. Stopper the flask with a solid rubber stopper and weigh it to the nearest 0.01 g. Add about 10 mL of NaOH solution to the flask, stopper it and weigh it to the nearest 0.01 g. If you measure the NaOH solution with a graduated cylinder make sure that it is clean and dry before using. Record the composition of the NaOH solution.

Obtain about 10 mL of vinegar in your 50-mL beaker. Use the eye dropper to add vinegar to the flask. Add the vinegar one dropper full at a time while you gently swirl the contents of the flask. Continue adding the vinegar until the color of the mixture changes to a yellow color. After adding the vinegar, stopper the flask and weigh it to the nearest 0.01 g. Be sure that the flask is not wet on the outside.

Use a dropper bottle of NaOH solution to add more NaOH to the flask, drop by drop. Carefully add drops of NaOH solution and swirl the contents after adding each drop. Continue adding drops until the mixture turns from yellow to a green or blue color. Take care not to add more drops than are needed to turn the mixture to green or blue. After adding the required number of drops, stopper the flask and weigh it to the nearest 0.01 g. Be sure that the flask is not wet on the outside. Repeat the experiment two or three more times so that you have three or four sets of data. You may use a second 50-mL flask and run two experiments concurrently.

Using your data calculate the mass of the vinegar sample, the mass of the 10- mL of NaOH solution, and the mass of NaOH solution drops. Add the two NaOH solution masses to obtain the total mass of NaOH solution used. Use these data to calculate the mass of acetic acid in your vinegar sample. Calculate the percent-by-mass of acetic acid in the vinegar by dividing the mass of acetic acid by the mass of the vinegar sample and multiplying by 100. If the results of your percents agree within 0.1, determine the average of the percents. If your percents do not seem to be close enough, consult your instructor.

11 REPORT SHEET EXPERIMENT 11

Name _____

Lab Section _____   Due Date _____

Composition of NaOH solution as provided _____

(A) Mass of flask +
    drop indicator   _____ _____ _____ _____

(B) Mass of flask +
    10 mL NaOH   _____ _____ _____ _____

(C) Mass of flask +
    vinegar added   _____ _____ _____ _____

(D) Mass of flask +
    drops NaOH   _____ _____ _____ _____

Mass 10 mL NaOH   _____ _____ _____ _____
    (B) - (A)

Mass NaOH drops   _____ _____ _____ _____
    (D) - (C)

Total mass NaOH   _____ _____ _____ _____

Mass vinegar   _____ _____ _____ _____
    (C) - (B)

Calculations: Show setup and answers.

   Mass of acetic acid in samples

   Percent-by-mass acetic acid in vinegar and average percent

11 QUESTIONS EXPERIMENT 11

Name KC Anderson

Lab Section _____ Due Date _____

1. Two brands of vinegar are analyzed by mixing samples with a sodium hydroxide solution containing 0.00267 moles NaOH/1 g solution. Using the following data, determine which of the brands has the higher percent-by-mass acetic acid.

	mass of vinegar sample	mass of NaOH solution needed
Brand X	35.76 g	19.41 g
Arlene's Deluxe Vinegar	27.60 g	14.92 g

2. When a vinegar solution is analyzed by mixing it with a sodium hydroxide solution of known composition, tell what effect each of the following factors has on the calculated percent of acetic acid. That is, would the calculated percent be greater than it should be, less than it should be, or not affected. Give an explanation of your answer. Note that the calculated percent is directly proportional to the mass of sodium hydroxide solution and inversely proportional to the mass of the vinegar.

(a) Too many drops of sodium hydroxide solution are added to the vinegar.

(b) Some vinegar solution dribbles down the outside of the flask while it is being added to the sodium hydroxide solution.

12 Energy in Chemistry

Objective

The purpose of this experiment is to observe energy exchanges that accompany changes of state and chemical reactions.

Terms to Know

Boiling Point - The temperature at which a liquid boils.

Solidification, Crystallization, or Freezing - The change from liquid to solid.

Freezing Point - The temperature at which a liquid freezes.

Melting Point - The temperature at which a solid melts.

Exothermic Reaction - A chemical reaction that releases energy.

Endothermic Reaction - A chemical reaction that absorbs energy.

Discussion

In normal environments, a chemical is either in the solid, liquid, or gaseous state.
Chemicals can be changed from one state to another by heating or cooling. For example, by heating ice (solid water) it can be made to melt, changed from the solid to the liquid state. By heating, liquid water can be boiled or vaporized, changed from the liquid state to the vapor or gaseous state. Under certain conditions some solids can be converted directly from the solid state to the gaseous state. You have probably observed this with a piece of dry ice (solid carbon dioxide). This change is called sublimation.

A gas can be cooled and caused to change to the liquid state. This is what happens when a window becomes fogged with water droplets that are formed from the water vapor in the air. This change is called condensation. A liquid can be cooled and changed to the solid state. This is, of course, what happens when water freezes to form ice. Such a change is called solidification, crystallization, or freezing. The temperature at which certain changes of state occur can be used to characterize and identify chemicals. The temperature at which a solid melts is called the melting point of the solid. For instance, ice melts at about 0 °C. Similarly, the temperature at which a liquid solidifies under specific conditions is called the freezing point of the liquid. For a pure chemical the freezing and melting points are the same. The temperature at which a liquid under specific conditions boils is called the boiling point of the liquid. The observation of freezing, melting, and boiling points is often carried out in the laboratory for purposes of describing chemicals.

When a liquid is heated it forms a vapor that exerts a pressure. A liquid at a specific temperature has a characteristic vapor pressure. The evaporation of a liquid exposed to the atmosphere is inhibited by the atmospheric pressure. As the liquid is heated, the vapor pressure will increase, and evaporation will occur more readily. When the temperature is raised so that the vapor pressure equals the atmospheric pressure, no inhibition of evaporation occurs and the liquid rapidly evaporates. The rapid evaporation under these conditions is called boiling. The temperature at which the vapor pressure of a liquid is equal to the atmospheric pressure is the boiling point of the liquid.

Energy exchanges are involved when chemicals undergo changes of state, such as melting, freezing, boiling, and condensing. The calorie and the joule are the common units used to express amounts of energy. Energy is required to melt a solid. When a liquefied solid is cooled, it will usually freeze at the same temperature at which the solid melted. This temperature is the freezing point of the chemical. When freezing occurs, energy is released. A summary of the energy exchanges that accompany changes of state is given below:

```
        --------------------------------------------------> energy absorbed

solid <--------------------> liquid  <----------------------> gas

energy released <--------------------------------------------------
```

Chemical reactions involve the breaking and forming of chemical bonds. When bonds break and are formed, energy exchanges are involved. Consequently, when a chemical reaction occurs, a corresponding energy exchange takes place. Some chemical reactions release heat to the surroundings and are called exothermic reactions. Other reactions take heat from the surroundings and are called endothermic reactions. Thus, it is sometimes possible to observe the energy exchange involving a reaction by observing whether heat is given off or absorbed. In a chemical reaction that is open to the atmosphere, the energy involved which takes the form of heat is called the enthalpy change or heat of reaction. The energy which is involved in a reaction can be shown in the equation to indicate an exothermic or endothermic reaction. For example, an exothermic reaction gives energy as a product

$$CH_4(g) + 2O_2(g) \longrightarrow CO_2(g) + 2H_2O(g) + energy$$

and an endothermic reaction shows energy as a reactant.

$$energy + N_2(g) + O_2(g) \longrightarrow 2NO(g)$$

When a chemical dissolves in water, chemical bonds may break and water molecules may be attracted to the various species formed upon dissolving. The energy exchange involved in dissolving is called the enthalpy or heat of solution. The heat of solution can be exothermic or endothermic depending upon the chemical which is dissolved. Is the dissolving process described by the following equation exothermic or endothermic?

$$energy + NaNO_2(s) \longrightarrow Na^+(aq) + NO_2^-(aq)$$

Some chemical reactions occur spontaneously when the reactants are mixed. However, in many cases, the reactants can be mixed and no reaction occurs. For example, we know that methane can react with oxygen in an exothermic reaction. However, when we turn the gas on in a Bunsen burner and allow the methane to mix with oxygen in the air, nothing happens. It takes a lighted match or spark to start the burning process. Many reactions behave in a similar manner. The reactants can be mixed and no observable reaction occurs. However, with some source of energy, such as a flame or spark, the reaction is initiated. For exothermic reactions the source of energy is just needed to start the reaction and, once it is started, it will proceed on its own as energy is released. On the other hand, endothermic reactions require a continuous source of energy to occur. An exothermic reaction can warm up the environment and an endothermic reaction can absorb heat from the environment.

Energy exchanges during some chemical reactions can be measured by carrying out the reaction in water surroundings and measuring the temperature change of the water. It requires 1 cal or 4.184 J of energy to change 1 g of water by 1 °C. This is called the specific heat of water. The mass of water and the temperature change is used with the specific heat of water to find the number of kilojoules of heat exchanged in a reaction involving a specific number of moles of reactants. This approach can be used to measure the heat of reaction.

Laboratory Procedure

1. THE BOILING POINT OF A LIQUID

Set up the apparatus shown in Fig. 12-1. The thermometer is suspended on an iron ring using a short length of copper wire. Be very careful not to break the thermometer. Place about 5 mL of the designated liquid in the test tube along with two or three boiling chips. The thermometer is immersed into the test tube so that the bulb is about 1 cm above the liquid surface but not touching the surface. Do not allow the thermometer to touch the sides of the test tube.

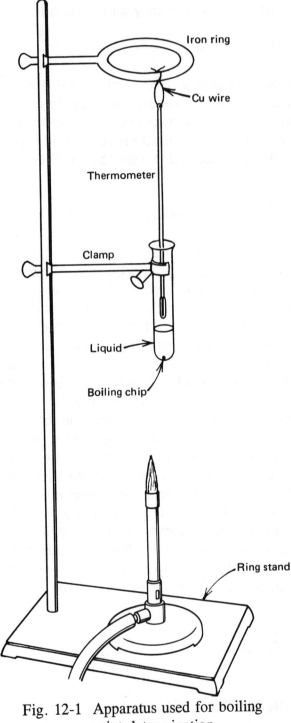

Fig. 12-1 Apparatus used for boiling
point determination

Light a Bunsen burner and, using your hand, raise it to heat the liquid in the test tube. Carefully heat the liquid until it begins to boil. Do not allow the liquid to boil out of the tube. Control the heating so that the liquid gently boils. As it boils, note that a drop forms on the bulb of the thermometer. When this occurs and the liquid is gently but continuously boiling, read the temperature registered by the thermometer. This is the boiling point of the liquid. Record your measured boiling point.

2. THE HEATING CURVE FOR WATER

In this section you are going to observe and measure the change in the temperature of water when it is heated. You will start with an ice water mixture and heat it until the water boils. Leave your thermometer suspended on the iron ring. Use the ring stand to set up a heating apparatus like that shown in Fig. A1-5. Be sure that the ring supporting the wire gauze is adjusted so that the hot part of the Bunsen burner flame will touch the gauze. Fill a 250-mL beaker with ice and fill it three quarters full of deionized water. Place the beaker on the wire gauze. Immerse the thermometer into the beaker of ice water and adjust it so that the bulb of the thermometer is about 1 cm from the bottom of the beaker. You will need a watch or clock with a second hand. You are going to heat the ice water and record the temperature every 30 seconds until the water boils.

Light your Bunsen burner and set it under the wire gauze supporting the beaker. Quickly read the temperature registered by the thermometer and record this as the temperature for time zero. Now read the thermometer and record the temperature every 30 seconds. Continue the heating and reading of the thermometer until the water boils vigorously and continue to read it for two minutes after vigorous boiling starts. The water should boil after about 10 minutes of heating.

Temperature

Time

Temperature

Time

To construct a heating curve for water plot the temperature versus the time in seconds on a piece of graph paper. Connect the points on the graph with a curved line. Be sure to label the axes on your graph and title the graph as "The Heating Curve for Water." See the back of the book for some graph paper and Experiment 6 for a discussion of graphing data.

3. HEAT OF REACTION

(a) Pour about 1 mL of 6 M hydrochloric acid into a clean test tube.(CAUTION: Acid and base solutions are DANGEROUS, so be careful not to spill or splash them.) Pour about 1 mL of 6 M sodium hydroxide solution into another tube. Now, carefully pour the acid into the sodium hydroxide solution. Feel the test tube and record your observations.

The following equation represents the reaction that occurred. Write in the word "energy" on the proper side of the equation to indicate whether energy was released or absorbed.

$$H_3O^+(aq) + OH^-(aq) \longrightarrow 2H_2O$$

(b) Pour about 2 mL of 6 M sodium hydroxide solution into a large test tube. Obtain four 1 cm squares of aluminum foil. Place a thermometer in the test tube and observe the temperature. Now drop the foil into the tube and observe the temperature as the reaction takes place. Give the reaction some time to start. Record your observations.

The following equation represents the reaction that occurred. Write the word "energy" on the proper side of the equation.

$$6H_2O + 2Al(s) + 2OH^-(aq) \longrightarrow 2Al(OH)_4^-(aq) + 3H_2(g)$$

(c) Pour about 0.5 mL of concentrated (18 M) sulfuric acid in a small test tube. (CAUTION: Concentrated H_2SO_4 is very dangerous. It is more concentrated and dangerous than battery acid.) Pour about 5 mL of water in a large test tube. Now, carefully pour the acid into the water, swirl gently, and cautiously feel the tube. Record your observations.

The following equation represents the reaction that occurred. Write the word "energy" on the proper side of the equation.

$$H_2SO_4(l) + H_2O \longrightarrow H_3O^+(aq) + HSO_4^-(aq)$$

Why should you always add sulfuric acid to water and never add water to sulfuric acid?

(d) Pour about 2 mL of 6 M hydrochloric acid into a small test tube. Obtain a small piece of magnesium metal. Place a thermometer in the test tube and observe the temperature. Now, drop the magnesium in the tube and observe the temperature as the reaction takes place. Record your observations.

The following equation represents the reaction that occurred. Write the word "energy" on the proper side of the equation.

$$2H_3O^+(aq) + Mg(s) \longrightarrow Mg^{2+}(aq) + H_2(g) + 2H_2O$$

(e) Put a sample of solid sodium bicarbonate into a test tube to have a sample that is about 0.5 cm deep in the tube. Pour about 2 mL of 1 M hydrochloric acid into the tube and feel the outside of the tube. Record your observations.

The following equation represents the reaction that occurred. Write the word "energy" on the proper side of the equation.

$$H_3O^+(aq) + NaHCO_3(s) \longrightarrow Na^+(aq) + CO_2(g) + 2H_2O$$

4. HEAT OF SOLUTION

(a) Place about 1 g of ammonium chloride, NH_4Cl, in a clean, dry test tube. Place a thermometer in the tube and observe the temperature. Pour in about 5 mL of water. Mix and observe any temperature change.

Write the word "energy" on the proper side of the equation representing the dissolving process.

$$NH_4Cl(s) \longrightarrow NH_4^+(aq) + Cl^-(aq)$$

(b) Pour about 2 mL of water in a large test tube. Obtain about 1 g of calcium chloride. Place a thermometer in the water and observe the temperature. Now, dump in the calcium chloride, then mix and observe the temperature as dissolving occurs. Record your observations below.

Write the word "energy" on the proper side of the equation representing the dissolving process.

$$CaCl_2(s) \longrightarrow Ca^{2+}(aq) + 2Cl^-(aq)$$

5. ESTIMATING THE HEAT OF REACTION

By observing the heat released in a chemical reaction that occurs in water surroundings it is possible to measure the heat of reaction. Such an experiment requires the precise measurement of temperature change. We can only obtain approximate values for the heat of a reaction using a typical laboratory thermometer.

The reaction needs to be carried out in an insulated vessel containing a known amount of water. Such an apparatus is called a calorimeter. We will use a calorimeter made of two styrofoam cups with a lid through which a stirring rod and a thermometer are immersed into the water contained in the cups. The calorimeter is supported in a 250-mL beaker. An example of an assembled calorimeter is on display in the laboratory.

Obtain the components of a calorimeter and assemble the parts so that you have a calorimeter like the example that is on display.

Heat of Reaction for Acetic Acid Reacting With Sodium Hydroxide

You are going to measure the approximate heat of reaction for the reaction represented as:

$$HC_2H_3O_2(aq) \ + \ NaOH(aq) \ \longrightarrow \ NaC_2H_3O_2(aq) \ + \ H_2O \ + \ energy$$

Use a clean, dry graduated cylinder to measure 50 mL of 1.0 M $HC_2H_3O_2$ as close to 50 mL as you can estimate. Remove the lid from the calorimeter and carefully pour all of the acetic acid solution into the styrofoam cup. Replace the lid and be sure that the thermometer is immersed in the liquid. Rinse and dry your graduated cylinder and measure 50 mL of 1.0 M NaOH as close to 50 mL as you can estimate. Read the temperature indicated by the thermometer in the calorimeter to the nearest 0.1 °C. Raise the lid of the calorimeter and carefully pour all of the sodium hydroxide solution into the calorimeter. Quickly replace the lid and gently stir the contents. Observe the temperature and note when it appears to stop rising. Record the highest temperature reading you notice to the nearest 0.1 °C. Pour the liquid from the calorimeter and rinse it with some deionized water. Gently dry the styrofoam cup, the lid, the thermometer and the stirring rod.

The heat of reaction can be expressed as the number of kilojoules of heat released per mole of $HC_2H_3O_2$. The number of kilojoules of heat released is found by multiplying the mass of water by the temperature change of the water and the specific heat of water. We can assume that there were 100 g of water in the calorimeter and we can use a specific heat of (4.184 J/g °C) for water. First we need to find the temperature change by subtracting the initial temperature from the final temperature.

$$\underline{\hspace{3cm}} - \underline{\hspace{3cm}} = \underline{\hspace{3cm}}$$

This can be used to calculate the number of kilojoules released:

(100 g) \underline{\hspace{2cm}} (4.184 J/g °C)(1 kJ/1000 J) = \underline{\hspace{3cm}} kJ

Since we know the molarity of the acetic acid used the number of moles of acetic acid that reacted can be found by multiplying the volume used by the molarity:

50 mL (1.0 mol $HC_2H_3O_2$/1000 mL) = \underline{\hspace{3cm}} mol $HC_2H_3O_2$

Finally the approximate heat of reaction is found by dividing the number of kilojoules of heat by the number of moles of acetic acid.

$$\underline{\hspace{4cm}} = \underline{\hspace{3cm}} \text{kJ/ mol } HC_2H_3O_2$$

12 REPORT SHEET EXPERIMENT 12

Name _____

Lab Section _____ Due Date _____

1. THE BOILING POINT OF A LIQUID

2. HEATING CURVE FOR WATER

Attach your plotted data to this report sheet.

3. Heat of Reaction

(a)

$$H_3O^+(aq) + OH^-(aq) \longrightarrow 2H_2O$$

(b)

$$6H_2O + 2Al(s) + 2OH^-(aq) \longrightarrow 2Al(OH)_4^-(aq) + 3H_2(g)$$

(c)

$$H_2SO_4(l) + H_2O \longrightarrow H_3O^+(aq + HSO_4^-(aq)$$

$$2H_3O^+(aq) + Mg(s) \longrightarrow Mg^{2+}(aq) + H_2(g) + 2H_2O$$

(e)

$$H_3O^+(aq) + NaHCO_3(s) \longrightarrow Na^+(aq) + CO_2(g) + 2H_2O$$

4. Heat of Solution

(a)

$$NH_4Cl(s) \longrightarrow NH_4^+(aq) + Cl^-(aq)$$

(b)

$$CaCl_2(s) \longrightarrow Ca^{2+}(aq) + 2Cl^-(aq)$$

5. Estimating the Heat of Reaction

Initial Temperature _____   Final Temperature _____

kilojoules of heat _____   moles of $HC_2H_3O_2$ _____

Estimated heat of reaction in kJ/mol $HC_2H_3O_2$ _____

12 QUESTIONS EXPERIMENT 12

Name _____

Lab Section _____ Due Date _____

1. Define the following terms:

 (a) Boiling Point

(b) Chemical Reaction

(c) Exothermic Reaction

(d) Endothermic Reaction

(e) Heat of Reaction

(f) Heat of Solution

2. Using the following experimental data, calculate the heat of reaction for the reaction of aluminum metal with hydrochloric acid in units of kilojoules per mole of aluminum. A 250 mL sample of 0.025 M hydrochloric acid is placed in a calorimeter and the initial temperature is found to be 20.6 °C. A 0.28 g sample of aluminum metal is added and the temperature rises to 25.9 °C. Assume that the calorimeter contains 250 mL of water. The reaction is:

$$2Al(s) + 6HCl(aq) \longrightarrow 2AlCl_3(aq) + 3H_2(g)$$

13 Light

Objective

To observe some examples of the relation between light and matter.

Terms to Know

Light or Electromagnetic Radiation (EMR) - Radiant energy that moves through space as waves.

White Light or Visible Light - EMR that is detectable by the human eye. Light that has a variety of colors including red, orange, yellow, green, blue, indigo and violet.

Emission Spectrum - The characteristic colors of light emitted by an element that has been energized.

Discharge Tube - A sealed glass tube with metal electrodes in which a sample of an element is placed. The element is subjected to electrical discharge and emits a characteristic pattern of light, its emission spectrum.

Discussion

Light and matter are related in an interesting way. If an object is heated to a very high temperature it will emit light. In fact the color of light emitted by a hot object reveals its temperature:

700 °C	dull red
900 °C	cherry red
1100 °C	orange
1300 °C	white

When a tungsten filament in an incandescent light bulb is electrically heated to a high temperature it gives off visible (white) light. Visible light is an example of a type of energy that radiates into the surroundings from its source. Visible light is a form of electromagnetic radiation, EMR. Other forms of electromagnetic radiation include radio waves, microwaves, infrared energy, ultraviolet energy, x rays and gamma rays.

Electromagnetic energy or radiant energy is energy passing through space in the form of waves. In other words, it is energy carried from its origin in the form of energy waves. You are most familiar with energy waves in water. When a pebble is dropped into a pool of water some of its energy moves outward in the form of energy waves. Waves in the ocean move energy from one location to another. Figure 13-1 shows a side view of a simple wave pattern.

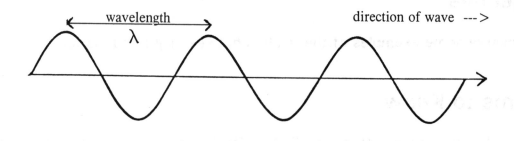

Fig. 13-1 A simple wave showing the wavelength

When a wave is momentarily fixed and viewed from the side it is pictured as a series of peaks and troughs. In real time the wave would roll in an undulating fashion. The wavelength of a wave is defined as the distance between adjacent peaks. Wavelength is sometimes represented by the Greek symbol lambda, λ . Ocean waves typically have wavelengths measured in meters. Various forms of electromagnetic radiation are described by their characteristic wavelengths. Wavelengths of electromagnetic radiation range from hundreds of meters to less than a trillionth of a meter. Often wavelengths of visible light are expressed in units of nanometers, nm. (1 nm = 1×10^{-9} m) Human eyes are sensitive to wavelengths of light between about 400 nm and 700 nm. This range includes all of the visible colors of light including red, orange, yellow, green, blue, indigo and violet. This sequence can be remembered using the mnemonic ROY G BIV.

Table 13-1 shows the varieties of radiant energy ordered by wavelength and frequency. A wave moves in a cyclic fashion. The distance spanned between peaks in a wave is one wavelength and the when a wave front travels one wavelength this is called one cycle of the wave. The frequency of a wave refers to the number of cycles that pass per second. Sometimes frequencies are expressed in terms of Hertz, Hz, which is a unit that represents one cycle per second. Shorter wavelength radiant energy is more frequent; has a higher frequency. Longer wavelength radiant energy is less frequent; has a lower frequency. The energy content of radiant energy is directly related to its frequency. Thus, higher frequency radiation is higher energy radiation and lower frequency radiation is lower energy radiation. As shown in Table 13-1 electromagnetic radiation ranges from long wavelength, low frequency and low energy radio waves to short wavelength, high frequency and high energy gamma radiation.

Name	Wavelength (meters)	Frequency (Hertz)	Comments
radio waves	10^3 to 1	3×10^5 to 3×10^8	AM 560 to 190 meters FM 3.4 to 2.8 meters TV 5.6 to 1.4 meters
microwaves	1 to 10^{-4}	3×10^8 to 3×10^{12}	Used for radar, communication and cooking.
infrared	10^{-4} to 7×10^{-7}	3×10^{12} to 4×10^{14}	Heat waves used in cooking. Next to red in the spectrum.
visible	7×10^{-7} to 4×10^{-7}	4×10^{14} to 7×10^{14}	Visible to human eye. ROY G BIV
ultraviolet	4×10^{-7} to 10^{-8}	7×10^{14} to 3×10^{16}	Causes sunburn. Higher (ultra) energy than violet
x rays	10^{-8} to 10^{-11}	3×10^{16} to 3×10^{19}	High energy and can penetrate the body.
gamma and cosmic rays	10^{-11} to 10^{-12}	3×10^{19} to 3×10^{20}	Very high energy and can penetrate the body.

Table 13-1 The Seven Varieties of EMR

Consider some of the ways in which matter and electromagnetic radiation interact. When a beam of radiant energy strikes matter it may be reflected, absorbed or scattered at various angles. A mirrored surface reflects visible light while a collection of tiny raindrops can scatter or diffract visible light revealing its various wavelengths as the colors of the rainbow. Chemicals in matter can absorb certain kinds of radiant energy. Some materials absorb infrared light which makes them warmer. Chemicals in our eyes absorb visible light which stimulates visual response in the brain. Ultraviolet light can be absorbed by chemicals in the skin and cause sun burn when some of the cellular chemicals are decomposed. X rays and gamma rays can be absorbed by cellular chemicals and decompose them. This is called cellular radiation damage. This damage can kill the cells and, in some cases, cause genetic damage to the cell. Sometimes genetic damage results in radiation induced cancer.

An opaque material reflects or scatter light and a transparent material transmits light. White light is polychromatic light composed of all wavelengths of visible light. White or clear materials reflect or transmit all wavelengths of visible light. A black material absorbs all wavelengths of visible light. A colored material contains chemicals that absorb various colors of visible light. The color reflected or transmitted by a material represents those wavelengths of light not absorbed by the chemicals. Thus, when a material appears to be distinctly red it is likely to be absorbing the colors and wavelengths of white light other than red. The color is a result of those wavelengths not absorbed.

Under certain conditions matter can emit electromagnetic radiation. When a material is heated to an elevated temperature it will emit radiation as infrared energy, sometimes called heat waves. When a substance is heated to a relatively high temperature it will emit visible light. A very hot object is said to be white hot. A white hot object, like the sun, also emits some infrared and ultraviolet light.

When visible light (white light) is dispersed or diffracted by tiny water droplets in the air it is split into its constituent colors; the colors of the rainbow. A diffraction grating is a device that can diffract visible light and separate it into its constituent colors. A source of white light viewed through a grating will appear as a rainbow since it will be diffracted into the various colors of the visible spectrum. A colored source of light, however, will reveal only specific parts of the visible spectrum corresponding to the color of the source.

A discharge tube is a specially designed cathode ray tube through which electricity flows under the influence of high voltage. The discharge tube is made of sealed glass in which a sample of an element has been added. It is possible to have a neon tube, a hydrogen tube, a sodium tube, a mercury tube, and so on. The cathode ray beam electrons interact with the atoms of elements in the tubes so that the atoms become energized. An energized atom has electrons that have higher energy than normal. As these energized electrons lose energy, they release characteristic wavelengths of light. Consequently the light given off by a discharge tube is characteristic of the element in the tube. A neon tube gives a distinct red light, a hydrogen tube gives a distinct red-purple light, a sodium tube gives a distinct yellow light, and, so on. When the light from a discharge tube is passed through a diffraction grating it is separated into wavelengths (specific spectral lines) that are characteristic of the element in the tube. The pattern of wavelengths that appear as a series of colored lines is called the emission spectrum of the element. Each element has a unique and characteristic emission spectrum. An element can be identified by its characteristic emission spectrum.

When compounds of some elements are placed in a flame characteristic light is emitted. Specific elements emit notable colors of light. In contrast some elements do not emit visible light. This behavior is the basis of identifying certain elements by flame tests. The colored light given off by fireworks comes from various compounds added to the explosives in the fireworks. The heat of the explosion excites electrons in the atoms of the elements and colored light results.

Laboratory Procedure

1. Light and Temperature

Light your Bunsen burner. Hold the tip of a stainless steel spatula in the hot part of the flame until it gets very hot. Remove the spatula and quickly note the color of the hot part. Describe how the color changes with time as it cools. Use the color to estimate the temperature of the hot spatula.

2. The Visible Light Spectrum

You need a diffraction grating or a spectroscope. Look at a source of visible light. (DANGER: If you use the sun as a source of light do not look directly at the sun.) Describe the colors of light you see in the spectrum and make a sketch with the colors labeled.

3. Absorption of Light by Colored Materials

You are going to observe the appearance of some colored plastics and the effect of overlapping various colors. Observe pieces of red, blue and yellow plastic sheets. If plastic sheets are not available you can make some colored solutions. Use three 100 mL beakers and fill each with 100 mL of water. Add and stir a drop of red food color to one beaker, a drop of yellow food color to the second beaker and a drop of blue food color to the third.

Observe the colored light transmitted by overlapping various colors of plastic. If you are using colored solutions hold the beakers next to one another with a source of light behind them. In each case describe the color and indicate which colors of visible light are being transmitted and which are being absorbed.

Red and yellow combination:

Blue and yellow combination:

Red and blue combination:

4. Observing the Light Absorbed by Chlorophyll

A sample of chlorophyll that has been extracted from a plant is available. Hold the vial of chlorophyll solution up to a source of light. Describe the color. What wavelengths or colors of visible light are likely not to be involved in photosynthesis?

5. Observing Light from Discharge Tubes

You are to observe the light from several different discharge tubes. First indicate the general color of light coming from a tube. Then, observe the emission spectrum of the light coming from a tube by holding a diffraction grating near the tube. Sketch the appearances of the emission spectra and indicate the colors of the specific lines.

Hydrogen Tube

Fluorescent Light Bulb

6. Absorption of Radiant Energy

Obtain a piece of paper that has small squares of aluminum foil and black paper attached. Hold the paper close to an incandescent light bulb for about one minute. Remove the paper and quickly touch the squares using two different fingers. Describe your observations and give an explanation.

7. Flame Tests

Solutions of compounds of a few elements are available in the lab along with a Bunsen burner. A cork with a wire or needle in it is provided for each solution. Dip the appropriate wire into a solution and place the tip of the wire in the edge of a Bunsen burner flame. Record your observations. Repeat the flame test for each solution. The metal present in each compound is responsible for the color. Record the metal name along with the corresponding color of the flame. A length of glass tubing is available near the solutions. Hold the tubing in the Bunsen flame and observe the color of the flame. Which of the elements you tested do you think is also present in the glass?

13 REPORT SHEET EXPERIMENT 13

Name _____

Lab Section _____   Due Date _____

1. Light and Temperature

2. The Visible Light Spectrum

3. Absorption of Light by Colored Materials

Red and yellow combination:

Blue and yellow combination:

Red and blue combination:

4. Observing the Light Absorbed by Chlorophyll

5. Observing Light from Discharge Tubes

Hydrogen Tube

Fluorescent Light Bulb

6. Absorption of Radiant Energy

7. Flame Tests

13 QUESTIONS EXPERIMENT 13

Name _____

Lab Section _____ Due Date _____

1. The heating elements on an electric stove glow orange-hot. What is the approximate temperature of the heating elements?

2. Fill in the following using the words higher and lower.

   Longer wavelength EMR is _____ energy.

   Shorter wavelength EMR is _____ energy.

   Higher frequency EMR is _____ energy.

   Lower frequency EMR is _____ energy.

3. What are the colors in a rainbow? In a normal rainbow is violet on the top of the bow or on the bottom? Why?

4. What is different about the light emitted from an incandescent bulb and the light emitted from a fluorescent bulb?

14 Lewis Dot Structure Worksheet

Name _____

Lab Section _____   Due Date _____

Write the Lewis electron dot structures for the following molecules and ions. The bonding sequence is given in some cases.

1. NF_3

2. SCl_2

3. $SOCl_2$ (central S)

4. CS_2

5. CCl_2F_2 (central C)

6. C_2H_6O (carbon-carbon-oxygen sequence)

7. C_2H_6O (carbon-oxygen-carbon sequence)

8. NH_2Cl (central nitrogen)

9. $HClO_2$ (central chlorine)

10. ClO^-

11. $PH_4{}^+$

12. C_2H_5Cl (carbon-carbon-chlorine sequence)

13. CNCOCN (carbon-carbon-carbon sequence)

14. CCl_3CHO (carbon-carbon sequence)

15 VSEPR   Worksheet

Name _____

Lab Section _____ Due Date _____

Use the VSEPR theory to predict the shapes of the following molecules and ions and sketch each shape. The central atoms are underlined and some of the molecules have more than one central atom. Predict the shape around each center.

1. $\underline{C}OF_2$

2. $\underline{C}F_4$

3. $\underline{P}Br_3$

4. $\underline{C}OS$

5. BrCl

6. $\underline{Se}Br_2$

7. $\underline{C}H_3\underline{C}O\underline{C}H_3$ (carbon-carbon-carbon sequence)

8. $\underline{C}H_3\underline{S}H$ (carbon center and sulfur center)

9. $Cl\underline{C}H_2\underline{C}N$ (carbon-carbon sequence)

10. $\underline{As}H_4^+$

11. $\underline{C}lO_2^-$

12. P_2O_3 (oxygen-phosphorus-oxygen-phosphorus-oxygen sequence)

16 Nomenclature Worksheet

Name _____

Lab Section _____ Due Date _____

1. Ionic Compounds Name the following ionic compounds and give the formulas of the ions that are contained in each compound.

	Name	Formulas of Ions
$Al(OH)_3$	_____	_____
Fe_2O_3	_____	_____
$(NH_4)_2SO_4$	_____	_____
$Hg(NO_3)_2$	_____	_____
$KC_2H_3O_2$	_____	_____
Na_3PO_4	_____	_____

2. Molecular Compounds Name the following molecular compounds.

Cl_2O_7 _____

BF_3 _____

P_4O_{10} _____

HI _____

3. Classify and name the following compounds:

Compound	Classification	Name
NH_3	_____	_____
$HC_2H_3O_2$	_____	_____
$FeSO_4$	_____	_____
CF_4	_____	_____
NO_2	_____	_____

4. Ionic Compounds Write the formulas of the following ionic compounds and give the formulas of the ions that are contained in each compound.

	Formula	Formulas of Ions
mercury(II) phosphate	_____	_____
zinc sulfide	_____	_____
barium nitrate	_____	_____
lithium carbonate	_____	_____
silver iodide	_____	_____
magnesium hydroxide	_____	_____

5. Molecular Compounds Write the formulas of the following molecular compounds.

diphosphorus pentoxide	_____
dinitrogen tetroxide	_____
hydrogen chloride	_____
oxalic acid	_____

6. Classify and give the formulas of the following compounds:

Compound	Classification	Formula
sodium cyanide	_____	
iron(III) nitrite	_____	
dibromine oxide	_____	
sulfuric acid	_____	
aluminum carbonate	_____	

17 The Detection of Common Ions

Objective

The purposes of this experiment are to perform chemical tests used for the identification of some common ions and to use the tests to detect ions in some familiar chemicals.

Terms to Know

Chemical Spot Test - a chemical reaction used to detect the presence of a specific element or ion.

Ammonium ion	NH_4^+
Chloride ion	Cl^-
Carbonate ion	CO_3^{2-}
Hydrogen carbonate ion	HCO_3^-
Sulfate ion	SO_4^{2-}
Nitrate ion	NO_3^-
Phosphate ion	PO_4^{3-}

Discussion

Chemical spot tests or qualitative tests are chemical reactions used to detect the presence of a specific element or ion. Spot tests are a form of qualitative analysis that reveal the presence of a chemical species in a mixture. This experiment includes some qualitative analysis tests for common ions.

Typically, a qualitative analysis test involves a chemical reaction that is unique for a specific ion. Selected chemicals are added to a solution to be tested for a given ion. If the ion is present, a chemical reaction occurs that gives a noticeable product. The unique reaction usually produces a solid product of notable color or form, or a gas that can be identified.

A qualitative analysis test is run by adding specific amounts of test chemicals to the solution or substance to be tested. If the unique reaction occurs, this indicates the presence of the ion. If no unique reaction occurs, the absence of the ion is indicated.

The common ions for which spot tests are given in this experiment are described below:

1. Ammonium ion, NH_4^+, is a common, positively charged polyatomic ion. Do not confuse the ammonium ion with the neutral compound ammonia, NH_3.

2. Chloride ion, Cl^-, is the very common simple ion formed by chlorine.

3. Carbonate ion, CO_3^{2-}, and hydrogen carbonate ion, HCO_3^-, are chemically related polyatomic ions containing carbon.

4. Sulfate ion, SO_4^{2-}, is a common sulfur-containing polyatomic ion.

5. Nitrate ion, NO_3^-, is the most common nitrogen-containing polyatomic ion.

6. Phosphate ion, PO_4^{3-}, is a common polyatomic ion of phosphorus.

Laboratory Procedure

1. QUALITATIVE ANALYSIS TESTS FOR COMMON IONS

The general directions for specific spot tests for common ions are given inside the following boxes. Use these directions any time you want to carry out a spot test for an ion. As you carry out the tests, observe and record any changes that occur when the chemicals are mixed. Be sure to use the amounts and volumes of chemicals and solutions as directed. Record your observations and indicate the results in each test. Each test is first run on a sample that is known to contain the ion. This allows you to see what a positive test looks like. Then a sample of an unknown is tested to see which of these ions it contains. Use small test tubes unless otherwise instructed. Do not follow the directions in the boxes until you are instructed to test for an ion.

Test for ammonium ion, NH_4^+:

1. Place 1 to 2 mL of solution to be tested in a test tube.

2. Add about 2 mL of a 6 M sodium hydroxide solution. (CAUTION:Caustic)

3. Hold moistened red litmus paper in the mouth of the test tube but do not allow the paper to touch the sides of the tube. It may be necessary to warm the tube, but do not let the solution boil. Do not breathe the fumes.

4. If the red litmus turns blue ammonium ion is indicated.

Test for chloride ion, Cl$^-$:

1. Place 1 to 2 mL of solution to be tested in a test tube.

2. Add 2 or 3 drops of 6 M nitric acid. (CAUTION: Acid)

3. Add 5 or 6 drops of 0.1 M silver nitrate solution.

4. If a distinct white solid forms chloride ion is indicated.

Test for carbonate ion, CO$_3^{2-}$, or hydrogen carbonate ion, HCO$_3^-$:

1. Place 1 or 2 mL of solution to be tested in a test tube.

2. Add 2 to 3 mL of 6 M sulfuric acid. (CAUTION: Acid)

3. Watch for the evidence of gas bubbles and test any gas being evolved by inserting a wire loop holding a drop of a barium hydroxide solution.

4. Note whether or not the barium hydroxide solution turns cloudy. If it turns cloudy carbonate ion or hydrogen carbonate ion is indicated.

Test for sulfate ion, SO$_4^{2-}$:

1. Place a small sample or 1 or 2 mL of solution to be tested in a test tube.

2. Add 2 or 3 drops of 6 M nitric acid. (CAUTION: Acid) Test with blue litmus to see if the solution the turns litmus red. If not add more acid.

3. Add 5 or 6 drops of 0.2 M barium chloride, BaCl$_2$, solution.

4. The formation of a white solid indicate that sulfate ion is indicated.

Test for nitrate ion, NO_3^- :

1. Place a small sample or 1 to 2 mL of solution to be tested in a large test tube.

2. Add 1 to 2 mL of 6 M sodium hydroxide, NaOH, solution. (CAUTION: Caustic) If ammonium ion or ammonia are known to be present, gently boil until red litmus no longer turns blue when inserted into the mouth of the tube. Be very careful not to spill any hot solution on yourself!

3. Add a 2 cm square of aluminum foil to the tube.

4. Push a small wad of cotton one half of the way into the tube to form a loose plug that prevents splashing of the solution.

5. Wipe the sides of the mouth of the tube to be sure that no solution is on the walls.

6. Bend a strip of red litmus paper into a V-shape, moisten with deionized water and insert it into the tube so that the tip of the V protrudes into the tube.

7. Gently heat the tube but do not boil. Heat until the aluminum dissolves.

8. If the litmus turns blue starting at the tip of the V nitrate ion is indicated.

Test for phosphate ion, PO_4^{3-} :

1. Place a small sample or 1 or 2 mL of the solution to be tested in a test tube.

2. Add 4 or 5 drops of 6 M nitric acid. (CAUTION: Acid)

3. Add 8 to 10 drops of an ammonium molybdate solution.

4. Gently warm the test tube but do not boil.

5. If a yellow solid forms phosphate ion is indicated.

THE CHEMISTRY INVOLVED IN THE SPOT TESTS

The chemical reactions that are part of each spot test are discussed in this section. For practice, perform the tests on solutions that contain specific ions as instructed. These tests will illustrate positive tests for these ions. Each time that you are instructed to test for a specific ion go to the appropriate box containing the spot test for that ion and follow the procedure.

A. AMMONIA, NH_3, AND AMMONIUM ION, NH_4^+ First, a solution will be tested for the presence of ammonia. Pour about 3 mL of household ammonia solution into a 50 mL beaker. CAUTION: Avoid breathing the fumes from the ammonia solution. Moisten a piece of red litmus paper with deionized water and hold it over the beaker so that it is close to, but not touching, the solution. Record your observations. The NH_3 gas contacting the litmus paper causes the color change.

Compounds containing the ammonium ion will not directly give a positive ammonia test. However, compounds or solutions containing ammonium ion can be treated with a sodium hydroxide solution to convert the ammonium ion to ammonia.

$$NH_4^+ + OH^- \longrightarrow NH_3 + H_2O$$

The ammonia formed can be detected by holding red litmus above the solution. The ammonia vapor will turn the red litmus blue.

USING THE AMMONIUM ION TEST ON SAMPLES:

Place about 1 mL of a 1 M ammonium chloride, NH_4Cl, solution in a small test tube and test for ammonium ion. Describe your results. You should obtain a positive test for NH_4^+.

B. CHLORIDE ION, Cl^- Chloride ion present in a soluble substance can be detected by dissolving the substance, adding nitric acid and then adding some silver nitrate, $AgNO_3$, solution. If chloride ion is present, a characteristic white solid is formed. This solid is silver chloride, $AgCl$, formed by the reaction:

$$Ag^+ + Cl^- \longrightarrow AgCl$$

USING THE CHLORIDE ION TEST ON SAMPLES:

a. Place a sample of table salt about the size of a pencil eraser in a test tube. Add 1 to 2 mL of deionized water and test for chloride ion. Record your observations. You should get a positive test for Cl^-.

b. Test a sample of tap water for chloride. Describe your results.

C. CARBONATE ION, CO_3^{2-}, OR HYDROGEN CARBONATE ION, HCO_3^-
Substances that contain carbonate ion or hydrogen carbonate ion react with acid

solutions to form carbon dioxide gas. The carbon dioxide gas usually bubbles out of solution. The reactions producing carbon dioxide are:

$$HCO_3^- + H_3O^+ \longrightarrow 2H_2O + CO_2$$

$$CO_3^{2-} + 2H_3O^+ \longrightarrow 3H_2O + CO_2$$

The presence of carbon dioxide can be confirmed as follows. A length of copper wire is bent to form a small loop on the end. The wire is dipped into a solution of barium hydroxide, $Ba(OH)_2$, to obtain a drop of solution in the loop. The wire in then carefully inserted into the tube in which the gas is produced. If the drop turns cloudy due to the formation of a white solid, the gas is carbon dioxide. The white solid is barium carbonate formed by the reaction:

$$Ba^{2+} + 2OH^- + CO_2 \longrightarrow BaCO_3 + H_2O$$

USING THE CARBONATE ION TEST ON SAMPLES:

Place a few pieces of marble in a test tube and test for carbonate ion. Pour the sulfuric acid onto the pieces of solid. Describe your results. You should get a positive test for carbonate ion.

D. SULFATE ION, SO_4^{2-} Sulfate ion present in a soluble substance can be detected by dissolving the substance, adding nitric acid and then adding some barium chloride, $BaCl_2$, solution. If sulfate ion is present, a characteristic white solid is formed. This solid is barium sulfate, $BaSO_4$, formed by the reaction:

$$Ba^{2+} + SO_4^{2-} \longrightarrow BaSO_4$$

USING THE SULFATE ION TEST ON SAMPLES:

Place about 1 mL of 0.1 M sodium sulfate, Na_2SO_4, in a test tube and test for sulfate ion. Record your observations. You should get a positive test for SO_4^{2-}.

E. NITRATE ION, NO_3^- It is possible to detect nitrate ion in soluble substances or solutions by adding sodium hydroxide solution and aluminum metal. A reaction occurs that changes the nitrate ion to ammonia. The reaction is:

$$3NO_3^- + 8Al + 5OH^- + 18H_2O \longrightarrow 8Al(OH)_4^- + 3NH_3$$

The ammonia can be detected using red litmus as was done in the ammonium ion test. Of course, ammonia and ammonium ion in a solution will interfere with the test for nitrate ion. They can be removed by adding NaOH solution and boiling the mixture.

USING THE NITRATE ION TEST ON SAMPLES:

Place 1 mL of a 1 M sodium nitrate, $NaNO_3$, solution in a test tube and test for nitrate ion. Record your observations. You should get a positive test for NO_3^-.

F. PHOSPHATE ION, PO_4^{3-} A test for phosphate ion in solution is accomplished by adding nitric acid followed by a solution of ammonium molybdate, $(NH_4)_6MoO_{24}$. If phosphate ion is present, a characteristic yellow solid of complex composition is formed.

USING THE PHOSPHATE ION TEST ON SAMPLES:

a. Place about 1 mL of a 0.1 M sodium phosphate solution, Na_3PO_4, in a test tube and test for phosphate ion. Record your observations. You should get a positive test for PO_4^{3-}.

b. Place a pencil-eraser size sample of a household detergent in a test tube and add 1 or 2 mL of deionized water. Test for phosphate ion. Record your results.

2. TESTING AN UNKNOWN

Your instructor will issue you a test tube of solution of a compound containing one of the following ions: NH_4^+, Cl^-, CO_3^{2-}, SO_4^{2-}, NO_3^-, or PO_4^{3-}. Record the number of the unknown. Test small portions of the solution for each of the ions. Be sure to use clean test tubes rinsed with deionized water. Record the identify of the unknown ion in your solution.

17 REPORT SHEET EXPERIMENT 17

Name _____

Lab Section _____   Due Date _____

1. Qualitative Tests for Common Ions

Record your observations of positive tests for each ion.

A. Ammonia, NH_3, and Ammonium ion, NH_4^+

B. Chloride ion, Cl^-

C. Carbonate ion, CO_3^{2-} or hydrogen carbonate ion, HCO_3^-

D. Sulfate ion, SO_4^{2-}

E. Nitrate ion, NO_3^-

F. Phosphate ion, PO_4^{3-}

2. Testing an Unknown Solution for an Ion

unknown number _____   Ion found _____

17 QUESTIONS EXPERIMENT 17

Name _____

Lab Section _____   Due Date _____

1. A compound is dissolved in water and tested for ions. A sample is mixed with a sodium hydroxide solution and litmus paper is held above the mixture. The litmus paper turns blue. Another sample is tested by adding nitric acid until the solution turns litmus paper red. Several drops of barium chloride solution are added and a white precipitate forms. Which ions are present in the compound and what is the formula of the compound?

2. Give the formulas for the compound ammonia and ammonium ion and explain the differences between the compound and the ion.

3. Give the names and formulas for common polyatomic ions containing:

   a. sulfur

   b. nitrogen

   c. carbon

   d. phosphorus

18 Gas Laws

Objectives

The purposes of this experiment are to:

1. Measure the volume of a gas sample at various pressures and a constant temperature.

2. Compare the measured volume of a gas with the volume calculated using Boyle's Law.

3. Measure the volume of a gas sample at various temperatures and a constant pressure.

4. Plot a graph of the volume of a gas sample versus the temperature.

Terms to Know

Boyle's Law - The volume of a gas at a constant temperature is inversely proportional to the pressure.

Charles' Law - The volume of a gas at a constant pressure is directly proportional to the temperature.

Percent Error - A way of expressing error in the experimental value of some quantity for which a theoretical value is known. The percent error is calculated by finding the difference between the experimental value and the theoretical value, dividing the difference by the theoretical value and multiplying the quotient by 100.

Discussion

This experiment is designed to allow you to confirm Boyle's Law and Charles' Law. Boyle's Law states the volume of a gas sample at a constant temperature is inversely proportional to the pressure. An algebraic expression of the law is $V = k/P$ where k is a proportionality constant. Boyle's Law means that if the pressure of a gas sample is increased, the volume will decrease and if the pressure is decreased, the volume will increase. This is an inverse proportion. Because of this proportionality, we can calculate the new volume of a gas sample if we know the initial volume and pressure and the final pressure of the gas. The new volume of a gas sample, V_f , which results when the pressure of the sample changes, is found by multiplying the initial volume, V_i, by a ratio of the pressure corresponding to the change.

$$V_f = V_i \frac{P_i}{P_f}$$

The pressure ratio should be greater than one if the initial pressure is decreased. This will give a new volume that is greater than the initial volume. The pressure ratio should be less than one if the initial pressure is increased. This will give a new volume that is less than the initial volume.

In this experiment you will measure the pressure and volume of a gas sample using a simple apparatus. It consists of a gas sample trapped between the sealed end of a narrow glass tube and a column of water that is free to move up and down the tubing. Figure 18-1 gives an illustration of the apparatus but it is not to scale. It is very important that you handle this tube with the utmost care. Handle it gently and do not shake it or move it abruptly. NOTE: If at any time the water column separates into parts or if any water comes out of the tube, notify your instructor. The tube is a delicate piece of equipment that will serve you well if you handle it carefully.

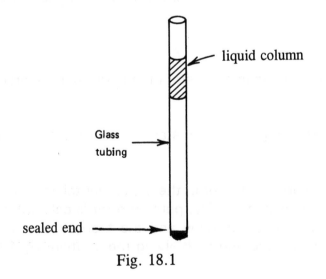

liquid column

Glass tubing

sealed end

Fig. 18.1

If the apparatus is in a horizontal position, the water column will adjust to a position in which the pressure of the gas in the tube equals the atmospheric pressure. That is, as shown in Figure 18-2, the pressures at points A, B and C will be equal to the atmospheric pressure.

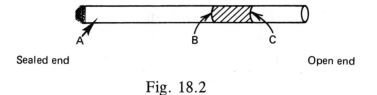

Sealed end   Open end

Fig. 18.2

When the tube is held vertically, as shown in Figure 18-3, the water column adjusts to a position in which the pressure of the gas balances the pressure of the atmosphere plus the additional pressure exerted by the water column. That is, the pressure of the gas equals the atmospheric pressure plus the pressure exerted by the column of water corresponding to the height of the water column between points B and C.

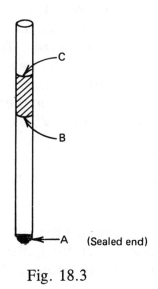

Fig. 18.3

When the tube is held with the open end pointing down, as shown in Figure 18-4, the pressure of the trapped gas is lower than the atmospheric pressure by the pressure corresponding to the height of the water column. That is, the pressure of the gas equals the atmospheric pressure minus the pressure corresponding to the height of the column of water between points B and C.

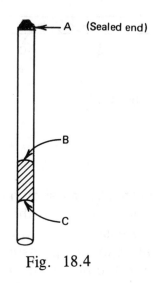

Fig. 18.4

Charles' Law states that the volume of a gas sample at a constant pressure is directly proportional to the temperature. An algebraic expression of the law is $V = kT$ where k is a proportionality constant. Charles' Law means that if the temperature of a gas sample is increased, the volume will increase and if the temperature is decreased, the volume will decrease. This is a direct proportion. Because of this proportionality, we can calculate the new volume of a gas sample if we know the initial volume and temperature and the final temperature of the gas. The new volume of the gas sample, V_f, which results when the temperature of the sample changes, is found by multiplying the initial volume, V_i, by a ratio of the temperatures corresponding to the change.

$$V_f = V_i \frac{T_f}{T_i}$$

The temperature ratio should be greater than one if the initial temperature is increased. This will give a new volume that is greater than the initial volume. The temperature ratio should be less than one if the initial temperature is decreased. This will give a new volume that is less than the initial volume.

In this experiment you will measure the volume and temperature of a gas sample using a short gas tube with a liquid plug. The volume of a trapped gas sample will be measured at various temperatures. The temperature is measured by attaching the liquid-plugged tube to a thermometer. Use two short lengths of rubber tubing or two rubber bands to hold the tube to the thermometer. One band near the bottom of the tube and one near the middle of the tube.

The gas sample in the tube has a cylindrical shape because this is the shape of the tube. The volume of the gas sample is the volume of the cylindrical space it occupies in the tube. It is not possible for us to measure this volume directly. However, the volume of the gas sample is directly related to the height of the sample from the bottom of the gas sample in the tube to the liquid plug. Consequently, we can measure the height of the sample and use it as a measure of the volume of the sample. However, we must measure the heights very carefully so that we have reliable measures of volume.

In this experiment you will study Boyle's Law by measuring the height of the gas sample at various pressures while the temperature is constant. Since the height of the gas sample is related to the volume of the sample, the height can be used as a measure of the volume. You will study Charles' Law by measuring the height of the gas sample at various temperatures while the pressure is constant. Again the height will serve as a measure of the volume of the gas sample.

Laboratory Procedure

A Reminder: The gas law tubes that you will use for this experiment is very fragile and must be handled with care. Do not shake them or handle them roughly. If the liquid plug separates, consult your instructor. You will make your height measurements with a metric rule. It is important to make these measurements as carefully as possible and try to estimate to the nearest 0.5 mm.

1. BOYLE'S LAW

Record the barometric pressure as provided. Obtain a long gas tube attached to a meter stick. The gas sample is contained by a column of water. The height of the column of gas in the tube changes as the pressure of the sample changes assuming the temperature is constant. The pressure of the sample is a combination of the atmospheric pressure and the pressure of the water plug. To express the pressure in torr corresponding to the water plug its length is measured and divided by 13.6. The 13.6 comes from the fact that the ratio of the density of mercury to the density of water is 13.6 to 1. The pressure of the gas sample is changed by holding the tube flat or with the open end up or down.

A. Volume at Atmospheric Pressure and Measuring the Water Column Length

Lay the tube and stick flat on the desk. Measure the length of the water column to the nearest 0.5 millimeter. Also carefully measure the length of gas sample from the sealed end to the position at which the gas meets the water column. To find the pressure corresponding to the water column divide the height of the water column by 13.6.

B. Volume at Atmospheric Pressure Plus the Water Column Length

Hold the tube and stick so the open end of the tube is pointing straight up. Measure the length of the gas sample from the sealed end to the position at which the gas meets the water column. The pressure of the sample is the atmospheric pressure plus the pressure corresponding to the water column. Calculate and record this pressure.

C. Volume at Atmospheric Pressure Minus the Water Column Length

Hold the tube and stick so the open end of the tube is pointing straight down. Measure the length of the gas sample from the sealed end to the position at which the gas meets the water column. The pressure of the sample is the atmospheric pressure minus the pressure corresponding to the water column. Calculate and record this pressure.

To see how your measured volumes correspond to Boyle's Law use the law to calculate the theoretical volumes and compare these to the measured volumes. Your measurements will have some experimental error and may not correspond exactly to the law. Remember that the height of the gas sample are an expression of sample volume. Use heights in your calculations. Use the atmospheric pressure as the initial pressure and the gas sample height in Part A as the initial volume. Calculate the theoretical volume of the gas sample when the pressure changes from atmospheric pressure to the pressure in part B. Compare the calculated result with the experimental height by determining the percent error. Find the percent error by subtracting the smaller value from the larger value, dividing by the value calculated using Boyle's law and multiplying by 100. Calculate the expected volume of the gas when the pressure changes from atmospheric pressure to the pressure in Part C. Determine the percent error between this value and the measured height in Part C.

Boyle's law can be expressed as $PV = k$. The compare your experimental measurements you can calculate the product of the pressure and height for each case. Theoretically the products should be the same. How close are they?

2. CHARLES' LAW

You need four 400-mL beakers, a thermometer, a metric ruler and a Charles' law gas tube. Put a length of magic tape vertically from the bottom to the top of one of the beakers. Allow a portion of the tape to overlap the top of the beaker so that it can be easily removed. Use rubber bands to attach the gas tube to the thermometer so that the bottom of the tube is flush with the bottom of the thermometer. Do not place a rubber band in the region of the bottom of the liquid plug in the tube or at the bottom of the tube. Support the tube and thermometer in a large beaker while you are not using it.

Fill another 400-mL beaker with water and heat it to a rapid boil. While the water is heating fill the third 400-mL beaker with room temperature water and the fourth beaker with ice and water. After the water begins to boil use a paper towel to pick up the beaker and pour the water into the 400-mL beaker with the tape on it. Hold the thermometer, with the gas tube attached, just above the surface of the hot water for a minute and then very slowly immerse it into the water. Be sure that the gas sample is completely immersed in the water but do not allow water to enter the open end of the tube. After about 2 or 3 minutes move the thermometer so that the gas tube is vertical and flush with the sides of the beaker. Adjust the gas tube so that you can see the gas sample along the edge of the tape. Carefully mark the position of the bottom of gas sample (not the bottom of the gas tube) and the position of the top of the gas sample where it meets the liquid plug. At the same time read and record the temperature to the nearest 0.1 degree. It is important to make the markings very carefully using a sharp pencil.

Remove the thermometer and gas tube from the beaker. Use a paper towel to hold the beaker of hot water and dump about one half of the hot water from the beaker. Fill the beaker with some room temperature water and stir to mix. Carefully immerse the thermometer and tube into the beaker. Hold the thermometer in the beaker for 2 or 3 minutes. Then, adjust the bottom of the gas sample to the previous mark on the tape and carefully mark the position of the top of the gas sample. At the same time read and record the temperature.

Once again remove the thermometer and tube. Dump half of the warm water from the beaker. Fill it with some room temperature water and stir to mix. Immerse the thermometer and tube into the beaker and hold it in place for 2 or 3 minutes. Then, adjust the bottom of the gas sample to the mark on the tape and carefully mark the position of the top of the gas sample. Read and record the temperature.

Remove the thermometer and tube from the beaker. Dump out the warm water and fill the beaker with tap water. Immerse the thermometer and tube into the beaker. Hold the thermometer in the beaker for 2 or 3 minutes. Then, adjust the bottom of the gas sample to the mark on the tape and carefully mark the position of the top of the gas sample. Read and record the temperature.

Remove the thermometer and tube from the beaker. Dump out the tap water and fill the beaker with the ice-water mixture. Immerse the thermometer and tube into the beaker. Hold the thermometer in the beaker for 3 or 4 minutes. Then, adjust the bottom of the gas sample to the mark on the tape and carefully mark the position of the top of the gas sample. Read and record the temperature.

Remove the gas tube from the thermometer and return it to the designated place. Use a metric rule to carefully measure the distances from the bottom mark on the

128

beaker to the other five marks on the tape. Make these measurements very carefully to the nearest 0.5 mm. Record the distances in the data table on the report sheet.

Charles' law is expressed as V = kT. This means that volume and temperature are directly proportional. A plot of volume versus temperature should give a straight line. On a piece of graph paper plot the heights of the gas sample in mm versus the temperature in degrees Celsius. After plotting the points use a straight edge to draw a straight line through the points. Use a line the overlaps or comes close to most of the points.

18 REPORT SHEET EXPERIMENT 18

Name _____ Lab Section _____

Due Date _____ `

Boyle's Law Data:

Atmospheric Pressure _____   Water Plug Length _____

Water Plug Pressure in torr _____   represented as WP below

	Height		Pressure
Horizontal	_____	P_{atm}	_____
Open up	_____	$P_{atm} + WP$	_____
Open down	_____	$P_{atm} - WP$	_____

Boyle's Law Calculations:

calculated volume _____   percent error _____

calculated volume _____   percent error _____

$PV = k$ calculations

Charles' Law Data:

Temperature	Height
_____	_____
_____	_____
_____	_____
_____	_____
_____	_____

Attach graph paper with Charles' Law plot to report sheet.

18 QUESTIONS EXPERIMENT 18

Name _____ Lab Section _____

Due Date _____

1. Explain why you could use the height of the column of gas as a measure of the volume of gas in the gas tube apparatus.

2. In the Boyle's Law experiment, why can you assume that the temperature was constant?

3. In the Charles' Law experiment, why can you assume that the pressure was constant?

4. To calculate with Charles' law temperatures need to be expressed in Kelvin. Make a table of your Charles' law data with heights and temperatures in Kelvin. Theoretically the ratio of V to T should be constant for each case: $V/T = k$. Calculate V/T for each of your experimental results.

19 The Molar Mass of a Gas

Objective

The purpose of this experiment is to determine the number of grams per mole of a gas by measuring the pressure, volume, temperature and mass of a sample.

Terms to Know

Molar Mass - The number of grams per mole of a substance.

Ideal Gas Law - The relation between pressure, volume, number of moles and temperature of a gas: $PV = nRT$

Vapor Pressure of Water - The pressure exerted by water vapor as it saturates the environment above a sample of liquid water. Water vapor pressures depend upon the temperature, and tabulated values are found in reference tables.

Discussion

The molar mass of a gaseous compound can be determined by experiment even though the formula or composition of the compound are not known. In other words, it is possible to find the molar mass of a gas even if we do not know its identity. The molar mass is determined by using the fact that the number of moles of a gas sample relates to the pressure, volume and temperature by the gas laws. The mass, pressure, volume and temperature of a gas sample are measured experimentally and used to calculate the molar mass.

To determine the molar mass of a gaseous substance from the mass, volume, temperature, and pressure of a sample, the number of moles in the sample is first calculated using the ideal gas law, PV = nRT. Recall that n represents the number of moles. The number of moles of a gas can be found from the pressure, volume and temperature of a sample: n = PV/RT where R is the universal gas constant.

As an example, consider a 0.508 g sample of a gas that occupies 522 mL volume at 100 °C and 0.960 atmospheres. First summarize the data as:

mass   P   V   T

0.508 g   0.960 atm   0.522 L   373 K (100 + 273)

The number of moles of gas is found using the ideal gas law. (Fill in the following blanks and calculate the number of moles.)

$$n = \frac{\underline{\hspace{3cm}} \text{ atm } \underline{\hspace{1cm}} \text{L}}{(0.0821 \text{ L atm/K mol}) \underline{\hspace{2cm}} \text{K}} = \underline{\hspace{2cm}} \text{ mol}$$

The molar mass of the gas is found from the ratio of the mass of the sample to the moles in the sample. Divide the mass of the sample by the number of moles using the value calculated above.

$$\frac{0.508 \text{ g}}{\underline{\hspace{2cm}} \text{ mol}} = \frac{\underline{\hspace{2cm}} \text{ g}}{1 \text{ mol}}$$

In this experiment a sample of gas will be collected by water displacement. That is, the gas will be bubbled into a container of water and as the gas accumulates, it will displace the water. The volume of the sample is the volume the gas occupies in the container. The temperature can be found by measuring the temperature of the water in contact with the gas. The mass of the gas sample is found by weighing a small tank or cylinder of gas, delivering the sample and reweighing the cylinder.

The pressure of the sample can be found by making sure that the pressure of the gas sample is the same as the atmospheric pressure. The atmospheric pressure can be measured with a barometer. However, when a gas is collected by water displacement, it becomes saturated with water vapor. This means that once the gas sample is collected, it will be a mixture of the gas and water vapor. This does not affect the volume or temperature of the gas but the measured pressure is the pressure of the mixture. That is, the measured pressure is the total pressure of the gas and the water vapor. To determine the pressure of the gas sample, the pressure of the water vapor can be subtracted from the total pressure. Appendix 2 lists the vapor pressure of water at various temperatures. To obtain the pressure of the gas sample in the experiment, look up the vapor pressure of water at the measured temperature and subtract this pressure from the measured atmospheric pressure:

$$P_{GAS} = P_{ATM} - P_{WATER}$$

Laboratory Procedure

For this experiment you will need a 800-mL beaker and a 250-mL Erlenmeyer flask. Place a piece of magic tape or a length of gummed label along the neck of the flask near the top. Fill the flask to the brim with tap water and put about 300 mL of water into the beaker. Allow the water to stand for a while to be sure that it is at room temperature. Set up the apparatus as shown in Figure 19-1 on the next page. Fill the beaker with about 300 mL of water. Stopper the flask with a rubber stopper and do not allow any air to become trapped in the flask under the stopper. Invert the flask through the iron ring which will hold it in position. Lower the ring with the flask so that the mouth of the flask is below the water level in the beaker. Secure the ring to support the flask. Use a spatula to remove the rubber stopper from the mouth of the flask. The stopper can remain in the beaker since it will not interfere with the experiment.

Obtain a small tank of gas. Use a paper towel to gently wipe the tank to make sure that it is clean. Weigh the tank to 0.01 g. Attach a plastic delivery tube to the valve on the tank and insert the glass tube on the other end into the mouth of the flask in the beaker of water. Push the valve on the tank to deliver gas to the flask. Continue to deliver gas until the flask is nearly full of gas. That is, deliver gas until the gas level is near the neck of the flask. Do not allow any gas to escape from the flask. If any gas escapes you will have to start the experiment over. Carefully remove the delivery tube from the gas tank, gently clean and dry the tank with a paper towel and weigh the tank to 0.01 g. Place a thermometer in the beaker.

Move the flask of gas so that the level of water in the flask is the same as the water level in the beaker. This will adjust the pressure in the flask to atmospheric pressure. Use a pen or pencil to place a mark on the tape to indicate the water level. Remove the flask from the water and fill it with water to the level corresponding to the mark on the tape. This volume of water will correspond to the volume of the gas sample. Carefully pour the water from the flask into a graduated cylinder and measure the volume to the nearest 1 mL. Record this as the volume of the gas sample.

Record the temperature of the water in the beaker. Assume that the temperature of the gas sample is the same as the water. Record the atmospheric pressure and look up the vapor pressure of water at the temperature of the gas in Appendix 2.

Repeat the experiment to obtain a second set of data. Calculate the molar mass of the gas using each set of data. If the results are within 10 percent of one another, express your answer as the average of the two results. If your two results seem to be too far apart consult your instructor.

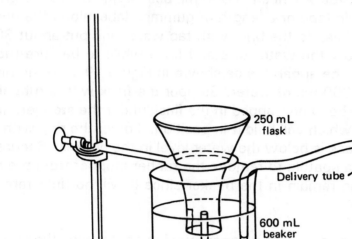

Fig. 19-1 Apparatus for collecting gas samples

19 REPORT SHEET EXPERIMENT 19

Name _____

Lab Section _____   Due Date _____

Mass of tank   _____ _____ _____

Mass of tank - sample   _____ _____ _____

Mass of sample   _____ _____ _____

Graduated cylinder volume   _____ _____ _____

Temperature of sample   _____ _____ _____

Atmospheric pressure   _____ _____ _____

Vapor pressure of water   _____ _____ _____

Pressure of sample   _____ _____ _____

Summary of Data

Mass of sample   _____ _____ _____

Volume in liters   _____ _____ _____

Temperature in Kelvin   _____ _____ _____

Pressure in atmospheres   _____ _____ _____

Give the setup and results of molar mass calculations below.

Average Molar Mass   _____

19 QUESTIONS EXPERIMENT 19

Name _____

Lab Section _____ Due Date _____

1. Referring to the experimental determination of the molar mass, explain how and why each of the following factors would affect your calculated result. That is, would the calculated result be greater than it should be, less than it should be, or not affected. Note that the relationships between the various factors involved in the calculation are:

$$g/mol = \frac{g}{\dfrac{PV}{RT}}$$

(a) The measured temperature is a lower value than the actual temperature.

(b) The measured volume is a higher value than the actual volume.

(c) Some of the gas sample escapes from the tank before it reaches the flask.

2. The molar mass of a gas is determined by collecting a gas sample by water displacement.

(a) Using the following data, calculate the molar mass of the gas: Sample volume, 163 mL; Sample temperature, 21 °C; Sample mass, 0.281 g; atmospheric pressure, 0.990 atm.

(b) If the gas in part (a) contains 85.5% C and 14.5% H, determine the empirical formula and use the result of part (a) to determine the actual formula.

20 Gas Laws Worksheet

Name _____

Lab Section _____ Due Date _____

1. A sample of a gas occupies a volume of 57.6 mL at 23 °C and 1 atm. What volume will it occupy at 296 K and 0.998 atm?

2. A sample of a gas occupies a volume of 121 mL at 26 °C and 767 torr. What volume will it occupy at 22 °C and 1.18 atm?

3. A gas sample is stored in a steel tank with a thermometer and a pressure gauge attached. Initially, the temperature reads 25 °C and the pressure gauge reads 23.5 atm. What will the pressure of the gas sample be when the gas is cooled to -38 °C?

4. A gas sample is stored in a flexible container and occupies a volume of 8.72 L at STP (Standard Temperature and Pressure is 0 °C and 1.00 atm.) What volume will the gas occupy at 746 torr and 297 K?

5. A gas sample occupies a volume of 17.8 L at a pressure of 735 torr and a temperature of 19 °C. If the sample is compressed so that it has a volume of 4.56 L at 19 °C, what is the new pressure of the sample?

6. A sample of neon gas occupies 24.5 L at 12 °C and 0.989 atm. How many moles of Ne are in the sample?

7. A sample of oxygen gas is saturated with water vapor so that it is a mixture of oxygen and water vapor. The sample occupies 396 mL volume at 21 °C and a total pressure of 771 torr. How many grams of oxygen are in the sample? Find the number of moles and then the mass. A table of the vapor pressures of water can be found in Appendix 2.

8. What volume does 1.00 mole of a gas occupy at 20 °C and 1.25 atm?

9. The density of a gas at STP is 3.14 g/L. What is the molar mass of this gas?

21 Solutions and Solubility

Objective

The purposes of this experiment are to investigate the nature of solutions, consider some factors that affect solubility, and to prepare a solution of known molarity.

Terms to Know

Solution - A homogeneous mixture of two or more components in which the component particles are intermixed on the molecular or ionic level.

Solvent - The component of a solution that acts as the dissolver of the other components.

Solute - The dissolved component of a solution.

Solubility - The amount of solute that will dissolve in a specified amount of solvent.

Saturated Solution - A solution in which dissolved and undissolved solute are in dynamic equilibrium. A solution that contains as much solute as can be dissolved at a specific temperature.

Supersaturated Solution - A solution that contains more solute than a saturated solution at the same conditions. Supersaturated solutions are unstable and readily revert to saturated solutions.

Molarity - The number of moles of solute per liter of solution: $M = n/V$

Percent-by-Mass Concentration - The number of grams of solute per 100 g of solution.

Discussion

Solutions are homogeneous mixtures in which the component particles are intermixed on the molecular or ionic level. When solid sugar is dissolved in water a liquid solution results. Water is the solvent. When a solution is prepared by mixing two chemicals of different physical states, the chemical that is of the same state as the resulting solution is called the solvent and the chemical that has been dissolved in the solvent is called the solute. If the two chemicals are of the same state, the solvent is usually considered to be the component present in the greatest amount. Generally, the solvent is the dissolver and the solute is the dissolved chemical.

Some solutes, like ammonia gas, are very soluble in water, and it is possible to prepare concentrated solutions of these solutes. Oxygen gas is somewhat soluble in water. Some solutes, like oxygen, are only slightly soluble in water, and some are so slightly soluble they are said to be insoluble. Many solid minerals are insoluble in water, however, some are soluble. That is why natural waters contain dissolved minerals. Generally, the solubilities of chemicals in water range from soluble to slightly soluble to insoluble. The solubility of a chemical in water is often expressed in terms of the grams of solute that will dissolve in 100 g of water at a specified temperature.

A saturated solution contains enough solute so that a state of dynamic equilibrium exists between the dissolved and undissolved solute. A solution that contains less solute than a saturated solution under given conditions is called an unsaturated solution. Such a solution will dissolve more solute if more is added. A supersaturated solution contains more solute than a saturated solution. Such a solution is unstable and the addition of a small crystal, or often merely mixing the liquid, will cause the excess solute to crystallize, leaving a saturated solution.

The composition of a solution is often stated in terms of a concentration that expresses the amount of solute dissolved in a specific amount of solution. A common concentration term is molarity, M, defined as the number of moles of solute per liter of solution.

$$\text{molarity} = \frac{\text{number of moles solute}}{\text{liter of solution}} = \frac{n}{V}$$

The molarity of a solution of known composition is calculated by dividing the number of moles of solute in the solution by the volume of the solution in liters. Percent by mass is another concentration term which expresses the number of grams of solute per 100 g of solution. The percent by mass solute of a solution of is calculated by dividing the number of grams of solute by the number of grams of solution and multiplying by 100.

Laboratory Procedure

Record your observations along with the answers to any questions.

1. SUPERSATURATED SOLUTION

Place 2 g of sodium thiosulfate pentahydrate, $Na_2S_2O_3 \cdot 5H_2O$, in a test tube, add five drops of deionized water. Heat without boiling until the solid is completely dissolved. Make sure it all dissolves, otherwise the exercise will fail. Place the test tube in the test tube rack and let it cool for about an hour. (Go on with other parts of the experiment and return to this part later.) After cooling, gently shake the test tube and observe the solution. If nothing happens, drop a small crystal of $Na_2S_2O_3 \cdot 5H_2O$ into the tube and observe. Feel the outside of the tube. Record your observations.

2. BOILING POINT ELEVATION AND FREEZING POINT DEPRESSION

The purpose of this part of the experiment is to observe how a solute affects the boiling point and freezing point of a solvent. Fill a large test tube one third full of solid sodium chloride. Set up a heating apparatus as shown in Figure A1-5. Place 100 mL of deionized water in a 250-mL beaker and heat it until it begins to boil. Hold a thermometer in the beaker and record the temperature of the boiling water. Do not leave the thermometer in the beaker without holding it. Pour the sodium chloride sample into the boiling water. Place the thermometer in the beaker of boiling water and note the temperature change.

Fill a large test tube one third full of solid sodium chloride. Obtain some crushed ice in a 100-mL beaker and add some deionized water to the beaker. Allow the ice water to sit for a few minutes. Hold a thermometer in the beaker and record the temperature of the ice water. Do not leave the thermometer in the beaker without holding it. Pour the sodium chloride sample into the ice water and stir with a stirring rod or spatula. Do not use a thermometer as a stirring rod. Note the temperature change.

3. RECRYSTALLIZATION

This part of the experiment is intended to illustrate how a chemical can be purified by dissolving it in a solvent and then allowing the chemical to recrystallize from the solution. Place 5 mL of deionized water in a large test tube and add 3 g of ammonium chloride, NH_4Cl. Heat the solution until all of the solid has dissolved but do not let the solution boil. After the solid has dissolved, cool the solution by holding the test tube in a stream of tap water. Describe your observations.

How could the process of recrystallization be carried out to remove an insoluble impurity? Hint: Insoluble solids can be removed from the liquid by filtration.

Why is a small amount of soluble impurity usually removed when the recrystallized solid is filtered out of the liquid in which it was dissolved and recrystallized?

4. SOLUBILITY OF AIR IN WATER

Water in contact with the atmosphere will dissolve slight amounts of oxygen and nitrogen gas from the air. Fill a 250-mL Erlenmeyer flask with tap water and insert a one-hole stopper. Make sure that the flask is full and that no air bubbles are trapped in the flask. Place an iron ring over the neck of the flask, hold a finger over the hole in the stopper, and invert the flask in a beaker of water. Support the flask with a ring so that the mouth is below the surface of the water in the flask. The apparatus is shown in Fig. 21-1. Carefully heat the water in the flask to near boiling, but DO NOT ALLOW IT TO BOIL. To avoid boiling, do not heat the flask at one point but move the flame around to heat the water evenly. As soon as you observe the release of the dissolved gases as tiny bubbles, stop the heating. Water does not decompose by this heating. Heating causes the dissolved oxygen and nitrogen gas to separated from solution as gases. Record your observations.

What do you conclude about the solubilities of these gases in water when the temperature rises?

Fill in the following table using the *Handbook of Chemistry and Physics*.

Solubility of	Cold Water	Temperature	Hot Water	Temperature
Oxygen, O_2				
Nitrogen, N_2				

What chemicals are present in the bubbles? Are they present in the same proportions as in the atmosphere? Why?

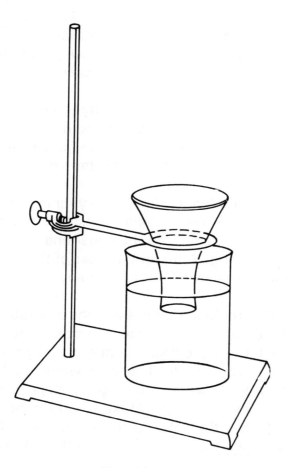

Fig. 21-1

5. THE CONCENTRATION OF A SOLUTION

The first step of this exercise involves the calibration of a flask to find the precise volume. You will need a clean, dry 125-mL Erlenmeyer flask, a rubber stopper for the flask, and a clean, dry 200-mL beaker. Carefully attach a gummed label or a piece of tape around the neck of the flask so that the top edge of the label forms a horizontal line slightly above the place where the neck meets the body of the flask. Stopper the flask and weigh it to the nearest 0.01 g. Obtain about 150-mL of deionized water in your 200-mL beaker. Pour the deionized water into the 125-mL flask and carefully fill it until the water level coincides with the top edge of the label marker. Make sure that the flask is dry on the outside. Stopper the flask and weigh it to the nearest 0.01 g. Pour the deionized water in the flask back into the beaker.

To find the mass of the water in the flask, subtract the mass of the empty flask from the mass of the flask filled with water. Using the density of 1.00 g/mL as the density of water, calculate the volume of water in the flask to three digits. Convert this volume from milliliters to liters to give the calibrated volume of the flask in liters.

Obtain a sample of sodium chloride, NaCl, from your instructor or use a about a 5 gram sample from the bottle provided. Weigh a piece of weighing paper to the nearest 0.01 g. Carefully place the sample on the paper and weigh the entire sodium chloride sample to the nearest 0.01 g. Subtract the mass of the paper to find the mass of the sodium chloride sample. Be careful not to spill any of the weighed sample.

Carefully pour the sodium chloride sample into the calibrated 125- mL flask and take care not to spill any of the sample. Using the deionized water in the beaker, pour about 50 mL of deionized water into the flask, stopper it and gently swirl to dissolve the sodium chloride. Now, carefully add deionized water until the solution level coincides with the top edge of the label marker. Tightly stopper the flask and gently shake the solution to make it homogeneous. Make sure that the flask is not wet on the outside. Weigh the flask and the solution to the nearest 0.01 g. Subtract the mass of the empty flask from the mass of the flask and the solution to find the mass of the solution.

Using the mass of the sodium chloride, the mass of the solution, and the volume of the solution (which is the same as the calibrated volume of the flask), calculate the molarity of the sodium chloride solution, the percent sodium chloride by mass in the solution, and the density of the sodium chloride solution. In the space provided on the report sheet, devise a label which describes the molarity of your solution.

21 REPORT SHEET EXPERIMENT 21

Name _____

Lab Section _____ Due Date _____

1. SUPERSATURATED SOLUTION

2. BOILING POINT ELEVATION AND FREEZING POINT DEPRESSION

3. RECRYSTALLIZATION

4. SOLUBILITY OF AIR IN WATER

Solubility of	Cold Water	Temperature	Hot Water	Temperature
Oxygen, O_2				
Nitrogen, N_2				

----

5. THE CONCENTRATION OF A SOLUTION

Mass of flask and water   _____

Mass of flask   _____

Mass of water in flask   _____

Volume of water in flask   _____   Volume in liters _____

Mass of NaCl and paper   _____   Mass of paper _____

Mass of NaCl sample   _____

Mass of flask and solution   _____

Mass of solution   _____

Data: Mass of NaCl sample   _____

     Mass of solution   _____

     Volume of solution   _____

Calculations (show setups and answers)

     Molarity of NaCl solution

     Percent NaCl by mass

     Density of NaCl solution

     Solution Label

21 QUESTIONS   EXPERIMENT 21

Name _____

Lab Section _____ Due Date _____

1. Most solutions we work with in the laboratory are (circle one)

    (a) unsaturated   (b) saturated   (c) supersaturated

2. The component of a solution sometimes designated as the dissolver is the (circle one)

    (a) solution   (b) solvent   (c) solute   (d) container

3. Some chemicals can be purified by removing impurities using (circle one)

    (a) dissolving   (b) recrystallization   (c) heating   (d) any of these

4. The solubility of air in water is decreased by (circle one)

    (a) heating   (b) cooling   (c) mixing   (d) any of these

5. A solution is prepared by dissolving 8.76 g of sodium sulfate, Na_2SO_4, in water to form 175 mL of solution. Calculate the molarity of the solution.

6. A water solution of glucose contains 49 g of glucose in 247 g of solution. Calculate the percent by mass of glucose in the solution.

22 Freezing Point Depression

Objective

In this exercise you are to use freezing point depression data in three ways.

1. measure the freezing point of pure water and use the data as a calibration of your thermometer.

2. prepare a water solution of ethyl alcohol and measure the freezing point of the solution. Use the data to estimate the freezing point depression constant for water.

3. prepare a water solution of a molecular solid and measure the freezing point of the solution. Use the data to estimate the molar mass of the solute.

4. determine the freezing point of a vinegar sample and use the data to estimate the percent by mass acetic acid in the vinegar.

Terms to Know

Freezing point - The temperature at which a liquid changes to a solid.

molality, m - The number of moles of solute per kilogram of solvent, (mol solute/kg solvent).

Freezing point depression - The difference between the freezing point of a pure solvent and the freezing point of a solution of the solvent, ΔT.

Freezing point depression equation - $\Delta T = K_f m$, where ΔT is the change in freezing point, m is the molality of the solute and K_f is the molal freezing point depression constant.

Discussion

The presence of a solute in a solvent makes the freezing point of the solution different from the freezing point of the pure solvent. Typically the freezing point of a solution is lower than that of the pure solvent. Freezing point depression data is used to approximate the molar mass of a molecular solute. This technique is sometimes called the cryoscopic method.

The change in the freezing point of a solvent is directly proportional to the amount of solute in the solution. If the amount of solute is expressed in terms of molality, m, the change in freezing point is given by the equation: $\Delta T = K_f m$ where ΔT is the change in freezing point, m is the molality of the solute and K_f is a proportionality constant called the molal freezing point depression constant. Each solvent has a unique value for this constant. For water the value of K_f is 1.86 °C/1 m. This means that the freezing point of water is depressed by 1.86 °C for every mole of solute dissolved in one kilogram of water.

The molality of a solution is the number of moles of solute per kilogram of solvent. The freezing point depression equation is put to various uses. For water solutions of known molality the equation is used to predict the freezing point depression, ΔT. An experimental value for the molal freezing point depression constant is found by measuring the freezing point depression for a solution of known molality and solving for the constant, $K_f = \Delta T/m$. The molality of a solute in a solution is found by measuring the freezing point depression of a solution, $m = \Delta T/K_f$.

Laboratory Procedure

Prepare an ice bath by filling a 400 mL or 600 mL beaker with ice. Add solid ammonium sulfate in abundance along with some water. Make sure that the bath is saturated with ammonium sulfate. Use a stirring rod to stir the ice-water mixture. As the ice melts add more. To have a low temperature bath you need to be sure that some of the ice melts and as it melts you need to add more. Stir the ice bath early and often. When you finish with your ice bath or you make a new one, please pour the ammonium sulfate solution into the collection vessel located in the lab. Since ammonium sulfate is a fertilizer the solution will be used to feed some deserving plants.

You need to know the quantitative composition of the solutions that you prepare so you can calculate the molalities. Prepare the solutions by measuring the mass of the solvent and the mass of the solute that is added to the solvent. Find the freezing point of water or a solution by placing about a 10 mL sample in a large test tube with a wire stirrer and a thermometer. Hold the test tube in an ice bath so that it cools. As it freezes continually stir the liquid to form a slush. If the liquid freezes on the edges of the tube without forming a slush remove the tube from the bath. Then rub the outside to melt some of the solid and stir vigorously to form a slush. Once you have a slush place the tube back into the bath and keep stirring it. The freezing point is the temperature at which solid and liquid are in contact. The temperature remains constant at the freezing point. Why?

Obtain a thermometer, some 18 gauge copper wire to make a stirrer, and four large, clean and dry test tubes.

A. Measure the freezing point of pure water as indicated by your thermometer. Read all temperature measurements as closely as possible to the nearest 0.1 °C. The freezing point may not be exactly 0 °C so read the temperature as closely as you can.

B. Weigh a clean-dry 50 mL flask to the nearest 0.01 g. Add about 50 g of water to the flask and weigh it. Use a clean-dry graduated cylinder to measure about 6 mL of ethyl alcohol. Add all of the ethyl alcohol to the flask of water and weigh the mixture to 0.01 g. Stopper the flask and shake to dissolve the solute. Find the freezing point of the solution to 0.1 ° C.

C. Weigh a clean-dry 50 mL flask to the nearest 0.01 g. Add about 50 g of water to the flask and weigh it. Use some weighing paper or a small beaker to carefully weigh about 4 grams of unknown solid to the nearest 0.01 g. Add all of the solid to the flask of water. Stopper the flask and shake to dissolve the solid. Find the freezing point of the solution to 0.1 ° C.

D. Pour about 10 mL of commercial white vinegar into a test tube. Find the freezing point of this solution to 0.1 ° C.

CALCULATIONS

The freezing point depressions for the solutions in parts B, C and D are found by finding the difference between the freezing point of pure water and the specific solution. Calculate the freezing point depression for each case.

B. Use the data for the ethyl alcohol solution (Part B above) to calculate the molality of the ethyl alcohol solution using the mass of ethyl alcohol, its molar mass and the number of kilograms of water in the solution. Use this molality and the freezing point depression of the ethyl alcohol solution in the freezing point depression equation to estimate the molal freezing point depression constant for water.

C. Use the freezing point depression equation, the known freezing point depression constant for water (1.86 ° C/m) and the freezing point depression of the unknown solute (Part C above) to find the molality of the solution of unknown. Multiply the molality by the number of kilograms of water in the solution to find the number of moles of unknown. Estimate the molar mass of the unknown solute by dividing the mass of unknown used in the solution by the number of moles.

152

D. Use the freezing point depression equation, the known freezing point depression constant for water (1.86 °C/m) and the freezing point depression of the vinegar (Part D above) to find the molality of the acetic acid in the vinegar solution. The molality has units of moles of acetic acid, $HC_2H_3O_2$, per 1000 grams of water. To find the percent acetic acid:

1. find the mass of acetic acid by multiplying the number of moles of acetic acid by its molar mass. Use the number of moles from the numerator of the molality.

2. find the total mass of solution containing this amount of acetic acid by adding the mass of acetic acid to 1000 g, which is the mass of water.

3. find the percent by mass of acetic acid by dividing the mass of acetic acid by the total mass of solution and multiplying by 100.

22 REPORT SHEET EXPERIMENT 22

Name _____

Lab Section _____ Due Date _____

Experimental Data

A. FREEZING POINT PURE WATER

B. ETHYL ALCOHOL SOLUTION _____

    mass of flask _____

    mass of flask + water _____

    mass of water _____

    mass of flask + water + alcohol _____

    mass of ethyl alcohol _____

Freezing point of alcohol solution _____

C. UNKNOWN SOLUTE

    mass of flask _____

    mass of flask + water _____

    mass of water _____

    mass of unknown solid _____

Freezing point of unknown solute solution _____

D. VINEGAR SOLUTION

Freezing point of vinegar _____

Calculations

Freezing point depression ethyl alcohol solution   _____

Freezing point depression unknown solute solution   _____

Freezing point depression vinegar   _____

B. ESTIMATE THE FREEZING POINT DEPRESSION CONSTANT FOR WATER

C. ESTIMATE THE MOLAR MASS OF THE UNKNOWN SOLUTE

D. FIND THE PERCENT BY MASS OF ACETIC ACID IN VINEGAR

22 QUESTIONS EXPERIMENT 22

Name _____

Lab Section _____ Due Date _____

1. A solution contains 20.0 g of sucrose dissolved in 100 g of water. What is the freezing point of the solution?

2. A solution contains 1.80 g of a molecular compound dissolved in 50.0 g of water. The freezing point depression for this solution is 0.372 °C. Calculate the molar mass of the compound.

3. A sample of vinegar has a freezing point depression of 2.0 °C. What is the estimated percent by mass of acetic acid in the vinegar?

4. Commercial hydrogen peroxide is 3.0 % by mass H_2O_2 in water. What is the predicted freezing point of this solution?

23 Conductivities of Solutions

Objectives

The purposes of this experiment are (1) to observe the conductivities of some solutions of chemicals and classify them as nonelectrolytes or strong or weak electrolytes, (2) to observe the changes in conductivity that take place when reactions involving ions occur, and (3) to observe the electrolysis of water. This exercise will be done as a demonstration in which you are to make observations as explained by your instructor.

Terms to Know

Electrolyte - A substance that forms an aqueous solution that conducts electricity.

Strong Electrolyte - An electrolyte that dissolves in water to form relatively high concentrations of ions.

Weak Electrolyte - An electrolyte that dissolves in water and reacts with water to a slight extent forming relatively low concentrations of ions.

Nonelectrolyte - A substance that forms an aqueous solution that does not conduct electricity.

Electrolysis of Water - The electrical decomposition of water into hydrogen gas and oxygen gas.

Discussion

Solutions containing ions can conduct electricity. If two metal wires are connected to the terminals of a battery (or generator) one wire can be considered to be negatively charged and the other positively charged. These wires are called electrodes and the negative electrode is the cathode and the positive electrode is the anode. When the electrodes are dipped into a solution containing ions, the positive ions (cations) are attracted to the cathode and the negative ions (anions) are attracted to the anode. The external battery can serve as an electron pump and, if the battery supplies a sufficient

157

driving force, chemical reactions involving the loss and gain of electrons occur at the electrodes. Electrons are gained by some species at the cathode and lost to form some other species at the anode. The migration of ions and the loss or gain of electrons provides a path for the flow of electrical current. Thus, the solution of ions is said to be conducting electricity. When a solution conducts electricity in this manner, a chemical is formed at the cathode and another at the anode. Such a process is called electrolysis.

By observing whether or not an aqueous solution of a chemical conducts, it is possible to tell if ions are present in the solution. A chemical that forms an aqueous solution that conducts electricity is called an electrolyte. A chemical that forms an aqueous solution that does not conduct is called a nonelectrolyte. Many molecular chemicals are nonelectrolytes. A nonelectrolyte does not form ions when it dissolves. Electrolytes form ions when they dissolve. Solutions of some electrolytes are strong conductors while solutions of others are weak conductors. The difference is due to the extent to which chemicals form ions when they are dissolved in water. Soluble ionic chemicals are strong electrolytes. In dissolving they break up or ionize into the constituent anion and cation. For example, when $CaCl_2$ dissolves in water it separates into ions:

$$CaCl_2(s) \longrightarrow Ca^{2+}(aq) + 2Cl^-(aq)$$

In general the dissolving of an ionic compound can be represented as:

$$C_nA_m(s) \longrightarrow nC^{m+}(aq) + mA^{n-}(aq)$$

Some molecular compounds are strong electrolytes since they react completely with water when dissolved to form cations and anions. Nitric acid is a strong electrolyte that reacts with water as shown by the following equation.

$$HNO_3 + H_2O \longrightarrow H_3O^+(aq) + NO_3^-(aq)$$

Some molecular chemicals react with water to a slight extent when dissolved to form relatively low concentrations of ions, but the major portion of the chemical remains in solution in the molecular form. These chemicals are weak electrolytes. Still other molecular chemicals dissolve in water and form no ions at all. These are the nonelectrolytes.

To represent the species in a solution of a chemical, it is standard practice to give the formulas of the major species present in the solution. Strong electrolytes are represented by the formulas of the ions they form in solution. Weak electrolytes and nonelectrolytes are represented by the molecular formula of the dissolved chemical. For example, a solution of calcium chloride contains Ca^{2+} and Cl^-, a solution of the weak electrolyte ammonia is represented as NH_3, and a solution of the nonelectrolyte hydrogen peroxide contains H_2O_2.

Laboratory Procedure

1. CONDUCTIVITY

Test the conductivities of the solutions and chemicals shown in the table in the report sheet and indicate whether they are good, weak or nonconductors. A typical conductivity apparatus is pictured in Fig. 23-1. After testing the conductivities, give the formulas of the major species in the various solutions. Note that hydrochloric acid (HCl in water) and sulfuric acid (H_2SO_4 in water) are similar to nitric acid mentioned in the discussion.

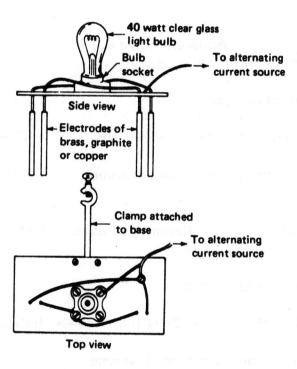

Fig. 23-1 Conductivity Apparatus

2. IONIC REACTIONS

Observe the conductivities of each of the following solutions. Then observe any change in conductivity that occurs when the solutions are mixed.

(a) Hydrochloric acid 0.1 M _____ Sodium hydroxide 0.1 M _____
 Change in conductivity as mixing occurs.

The equation for the chemical reaction is:

$$H_3O^+(aq) + Cl^-(aq) + Na^+(aq) + OH^-(aq) \longrightarrow Na^+(aq) + Cl^-(aq) + 2H_2O$$

Explain why any change in conductivity was observed.

(b) Acetic acid 0.1 M _____ Aqua ammonia 0.1 M _____
 Change in conductivity as mixing occurs.

The equation for the chemical reaction is:

$$HC_2H_3O_2(aq) + NH_3(aq) \longrightarrow NH_4^+(aq) + C_2H_3O_2^-(aq)$$

Explain why any change in conductivity was observed.

(c) Copper(II) sulfate 0.1 M _____ Barium hydroxide 0.1 M _____
 Change in conductivity as mixing occurs.

The equation for the chemical reaction is:

$$Cu^{2+}(aq) + SO_4^{2-}(aq) + Ba^{2+}(aq) + 2OH^-(aq) \longrightarrow BaSO_4(s) + Cu(OH)_2(s)$$

Explain why any change in conductivity was observed.

(d) CaCO₃ (marble) chips in water _____ acetic acid 6 M _____
 Change in conductivity as mixing occurs.

The equation for the chemical reaction is:

$$CaCO_3(s) + 2HC_2H_3O_2(aq) \longrightarrow Ca^{2+}(aq) + 2C_2H_3O_2^-(aq) + CO_2(g) + H_2O$$

Explain why any change in conductivity was observed.

3. ELECTROLYSIS OF WATER (Optional)

Water can be decomposed into hydrogen and oxygen by the process of electrolysis using a special apparatus in which electricity is passed through a water sample. (See Fig. 23-2.)   Give the balanced equation for the decomposition of water.

Observe the electrolysis of water and describe your observations.

What relative proportions of hydrogen and oxygen are produced during electrolysis?

Explain how you can deduce the empirical formula of water based on your observations of the electrolysis of water.

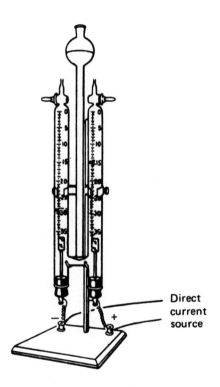

Fig. 23-2 Electrolysis Apparatus

23 REPORT SHEET EXPERIMENT 23

Name _____

Lab Section _____   Due Date _____

1. Conductivity

Solution or Chemical	Conductivity and Formula or Formulas of Major Species
Pure (deionized) water	_____
Tap water	_____
Acetic acid, 18 M	_____
Acetic acid, 0.1 M	_____
Aqua ammonia, 0.1 M	_____
Ethyl alcohol (pure)	_____
Ethyl alcohol solution	_____
Hydrochloric acid, 12 M	_____
Hydrochloric acid, 0.1 M	_____
Hydrogen peroxide solution	_____
Sodium acetate, 0.1 M	_____
Sodium chloride (solid)	_____
Sodium chloride, 0.1 M	_____
Sodium hydroxide, 0.1 M	_____
Sulfuric acid, concentrated	_____
Sulfuric acid, 0.1 M	_____
Sucrose (solid)	_____
Sucrose solution	_____

2. Ionic Reactions

a. Hydrochloric acid 0.1 M _____ Sodium hydroxide 0.1 M _____

b. Acetic acid 0.1 M _____ Aqua ammonia 0.1 M _____

c. Copper(II) sulfate 0.1 M _____ Barium hydroxide 0.1 M _____

d. $CaCO_3$ (marble) chips in water _____ Acetic acid 6M _____

3. Electrolysis of water

23 QUESTIONS EXPERIMENT 23

Name _____

Lab Section _____   Due Date _____

1. Define the following terms:

   (a) Electrolyte

   (b) Weak Electrolyte

2. Predict the formula or formulas of the major species contained in solutions of the following compounds given the conductivity behavior.

(a) A solution of $MgCl_2$ a strong conductor.

(b) A solution of sulfuric acid, H_2SO_4, is a strong conductor like hydrochloric acid.

(c) A solution of NH_4NO_3 is a strong conductor.

(d) A solution of sucrose, $C_{12}H_{22}O_{11}$, is a nonconductor.

(e) A solution of phosphoric acid, H_3PO_4, is a weak conductor.

(f) A solution of antifreeze, ethylene glycol, $C_2H_6O_2$, is a nonconductor.

24 Solution Concentration Worksheet

Name _____

Lab Section _____ Due Date _____

1. A solution contains 25.0 g of KNO_3 dissolved in water to make 250 mL of solution. Calculate the molarity of potassium nitrate in the solution.

2. How many moles of dissolved NaCl are in 25.0 mL of a 0.500 M NaCl solution?

3. How many grams of NaCl are contained in a 100 mL sample of a 0.500 M NaCl solution?

4. How many grams of LiCl are needed to prepare 500 mL of a 0.100 M LiCl solution?

5. What volume of a 0.100 M NaOH solution is needed to provide 0.0250 moles dissolved NaOH?

6. A solution contains 25.0 g of glucose dissolved in 85.0 g of water. What is the percent by mass glucose in the solution?

7. A solution is 10% by mass NaCl. How many grams of this solution are needed to provide 50 g of NaCl?

8. A commercial rubbing alcohol solution contains 70 g of isopropyl alcohol, C_3H_8O, dissolved in 30 g of water. What is the molality of isopropyl alcohol in this solution?

9. A solution has a molality of 0.91 m H_2O_2. What is the percent by mass H_2O_2 in this solution?

10. A solution of dilute nitric acid contains 19% percent by mass dissolved HNO_3 and has a density of 1.11 g/mL.   What is the molarity of the HNO_3 in this solution?

25 Chemical Reactions

Objective

In this exercise you are to observe

1. some chemical reactions and write net-ionic equations for them.

2. how the rate of a reaction is affected by concentration, state of subdivision and temperature.

3. the catalytic affect of some chemicals on the decomposition of hydrogen peroxide.

Terms to Know

Chemical Reaction - A process in which one set of chemicals changes into another set of chemicals.

Net-Ionic Equation - An equation for a reaction that excludes ions that do not react and includes only those compounds and ions that are reactants or products.

Reaction Rate - The speed at which a chemical reaction occurs.

Catalyst - A chemical that increases the rate of a reaction but is not chemically changed in the process.

Discussion

For chemical reactions that involve chemicals in solution it is possible to write overall equations or net-ionic equations. An overall equation shows the complete formulas of all compounds involved in the reaction. The net-ionic equation excludes those ions that are not reactants or products. These ions are called spectator ions since they are present in the reaction mixture but are not chemically changed in the reaction. In the first part of this exercise you are going to observe some chemical reactions and write the overall and net-ionic equations.

The rates of chemical reactions vary greatly. Some reactions are very fast and occur in seconds or fractions of seconds. In contrast, other reactions require minutes, hours or even longer periods of time. The major factors that effect reaction rates are concentration of reactants, state of subdivision, temperature and catalysts. In the second part of this exercise you are going to observe the effect of concentration, state of subdivision and temperature on the reaction of hydrochloric acid with calcium carbonate in Tums tablets. In the third part of this exercise you are going to observe the catalytic effect of various chemicals on the rate of the reaction involving the decomposition of hydrogen peroxide.

Laboratory Procedure

1. Observing Reactions and Writing Equations

For each of the following cases record any evidence that a chemical reaction has occurred and write the overall equation and net-ionic equation representing the reaction.

A. Place a 1 cm square of aluminum foil in a test tube. Add 2 mL of 6 M HCl to the tube. (CAUTION: 6 M hydrochloric acid is dangerous.) Record your observations. Translate the following overall equation into a net-ionic equation. Hydrochloric acid contains H_3O^+ and Cl^- and water is a product in the net-ionic equation.

$$2Al(s) \; + \; 6HCl(aq) \; \longrightarrow \; 2AlCl_3(aq) \; + \; 3H_2(g)$$

B. Place a few small pieces of solid calcium carbonate (marble chips) in a test tube. Add about 1 mL of 6 M HCl. (CAUTION: Acid) Record your observations. The reactants are calcium carbonate and hydrochloric acid. The products are an aqueous solution of calcium chloride, carbon dioxide gas and water. Hydrochloric acid contains H_3O^+ and Cl^-. Give the overall and net-ionic equations.

C. Place a 1 cm square of aluminum foil in a test tube. Add 1 mL of 0.1 M $Cu(NO_3)_2$ to the tube. Record your observations. The reactants are solid aluminum metal and an aqueous solution of copper(II) nitrate. The products are an aqueous solution of aluminum nitrate and solid copper metal. Give the overall and net-ionic equations.

D. Place 1 mL of 0.1 M $Cu(NO_3)_2$ in a test tube. Add 1 mL of 1 M NH_3 to the tube. Record your observations. The reactants are an aqueous solution of copper(II) nitrate and ammonia. The product is an aqueous solution of $[Cu(NH_3)_4](NO_3)_2$ which includes the complex ion $Cu(NH_3)_4^{2+}$ and the nitrate ion. Give the overall and net-ionic equations.

E. Work in a fume hood. Place a coiled 10-cm length of 24 gauge copper wire in a test tube. Add 2 mL of 6 M HNO_3 to the tube. (CAUTION: 6 M nitric acid is dangerous.) Record your observations and rinse the tube with water while working in the fume hood. The reactants are solid copper metal and aqueous nitric acid, $HNO_3(aq)$. The products are aqueous copper(II) nitrate, gaseous nitrogen dioxide and liquid water. Nitric acid solution contains H_3O^+ and NO_3^-. Give the overall and net-ionic equations. (Hint: When writing the net-ionic equation first balance the charge and then the nitrogens and add water as a product.)

2. Rates of Reactions

Concentration

You need three larger (1.8 cm x 15 cm) test tubes and a watch or clock for timing. Obtain three Tums tablets. Place one test tube in each of three 250-mL beakers. Add 20 mL of 1 M HCl to one tube, 20 mL of 2 M HCl to the second tube and 20 mL of 3 M HCl to the third tube. Drop one Tums tablet in each test tube and start timing. As you observe a tube you can gently swirl it. Record the time at which bubbles stop forming in each case.

hydrochloric acid	time
1 M HCl	
2 M HCl	
3 M HCl	

State of Subdivision

Obtain three Tums tablets. Place one (1.8 cm x 15 cm) test tube in each of three 250-mL beakers and add 20 mL of 2 M HCl to each tube. Place the whole tablet in one tube and start timing. Gently swirl the tube and record the time at which bubbles stop forming. Place a tablet on a creased clean sheet of paper and break it into eight or ten pieces of somewhat equal size. A metal spatula may be used to help break the tablet. Use the creased paper to dump the pieces into the second tube and start timing. Gently swirl the tube and record the time at which the reaction appears to stop. Crease a clean sheet of paper and fold it over a tablet. Crush the tablet and use a spatula to be sure that the pieces do not stick the paper. Use the creased paper to pour the tablet into the third tube and start timing. Gently swirl the tube and record the time at which the reaction appears to stop.

	time
whole tablet	
eight to ten pieces	
crushed tablet	

Temperature

Obtain three Tums tablets and a thermometer. Place one (1.8 cm x 15 cm) test tube in each of three 250-mL beakers. Add 20 mL of 2 M HCl to each tube. Fill one beaker with ice and add water to make an ice bath. Place one test tube in the ice bath. Fill the second beaker with tap water and place the second test tube in this beaker. Boil some water and add boiling water to the third beaker. Place the third test tube in this beaker and allow it to sit for 5 minutes. Place a thermometer in the hot water beaker. Drop one tablet into each of the three test tubes and start timing. Record the time that the reaction appears to stop in each case and record the temperature. Move the thermometer to the room temperature beaker and, then, to the ice bath as the reactions stop.

	time	temperature
hot water		
room temperature water		
ice bath water		

3. Catalysts

Hydrogen peroxide can decompose to give oxygen and water.

$$2H_2O_2 \longrightarrow 2H_2O + O_2$$

The reaction is noticeable when the oxygen gas bubbles from the solution. However, the decomposition of hydrogen peroxide in a water solution is not noticeable at room temperature. In this section, several chemicals are to be tested for their catalytic effect on the decomposition of hydrogen peroxide. Place seven small test tubes in a test tube rack and put into separate tubes a small amount (about the size of a match head) of each of the chemicals listed in the table below and for the solution add 10 drops. Label the tubes or rack to note which test tube has which chemical.

Obtain about 10 mL of a fresh 3% hydrogen peroxide solution. Gently heat the tube in the flame but do not boil. Record any evidence of decomposition as represented by gas bubbles. Pour about 1 mL of the heated hydrogen peroxide solution into each of the tubes in the rack. The rate at which bubbles of oxygen are formed in a test tube can be used as a measure of the catalytic effect. Observe and describe the catalytic effect or lack of effect of each chemical. An effect may be described as none, very slight, slight, great or very great. Record your results in the following table.

Chemical Tested	Catalytic Effect
salt, NaCl	
1 M $FeCl_3$ solution	
iron (III) oxide, Fe_2O_3	
iron metal (steel wool)	
marble, $CaCO_3$	
crushed piece of a green leaf	
yeast granules	

Why is it dangerous to store hydrogen peroxide solutions in bottles with iron lids?

Why is yeast such a good catalyst for the decomposition of hydrogen peroxide?

25 REPORT SHEET EXPERIMENT 25

Name _____

Lab Section _____ Due Date _____

1. Observing Reactions and Writing Equations

A.

B.

C.

D.

E.

2. Rates of Reactions

Concentration

hydrochloric acid	time
1 M HCl	
2 M HCl	
3 M HCl	

State of Subdivision

	time
whole tablet	
eight to ten pieces	
crushed tablet	

Temperature

	time	temperature
hot water		
room temperature water		
ice bath water		

3. Catalysts

Chemical Tested	**Catalytic Effect**
salt, NaCl	
1 M $FeCl_3$ solution	
iron (III) oxide, Fe_2O_3	
iron metal (steel wool)	
marble, $CaCO_3$	
crushed piece of a green leaf	
yeast granules	

Why is it dangerous to store hydrogen peroxide solutions in bottles with iron lids?

Why is yeast such a good catalyst for the decomposition of hydrogen peroxide?

25 QUESTIONS EXPERIMENT 25

Name _____

Lab Section _____ Due Date _____

1. Give the overall and net-ionic equations for the following reactions.

(a) Solid magnesium metal reacts in a 6 M HCl solution to give an aqueous solution of magnesium chloride and hydrogen gas. (Hint: Water is a product in the net-ionic equation.)

(b) An aqueous solution of formic acid, HCO_2H, reacts with an aqueous solution of NaOH to give an aqueous solution of sodium formate, $NaHCO_2$, and water.

2. In each of the following cases, explain what happens in terms of the factor or factors that affect reaction rates.

(a) Accumulated coal dust in a coal mine causes an explosion.

(b) As the small pieces of wood in a camp fire burn, the fire dies out and smolders.

(c) A football player feels temporarily energized after breathing pure oxygen.

(d) To store a battery keep it in the refrigerator but do not freeze it.

3. What is a catalyst? Give two examples of catalysts.

26 Precipitation Reactions

Objective

The purpose of this experiment is to carry out some precipitation reactions, observe them and to write the net-ionic equations for the reactions.

Terms to Know

Ionic Compound - A compound composed cations and anions.

Cation - An ion that carries a positive charge. Common cations are metal ions.

Anion - An ion that carries a negative charge. Common anions are polyatomic ions and simple nonmetal ions.

Soluble - A compound that can dissolve in water is said to be soluble. A compound is considered to be soluble if it will dissolve in water to make a solution having a concentration that is 0.1 M or greater.

Precipitation Reaction - A chemical reaction in which cations ions and anions in water solution combine to form an insoluble ionic compound. An ion combination reaction.

Precipitate - An insoluble solid formed in a precipitation reaction.

Solubility List - A list or table that summarizes the solubilities of common ionic compounds. A solubility list is based upon experimental observations of solubilities.

Discussion

When an ionic compound dissolves in water it separates into cations and anions. Thus, soluble ionic compounds exist in solution in the form of aqueous ions. Ions in solution are in continuous motion and migrate about the solution momentarily interacting with one another. As long as the compound corresponding to the cation and anion is soluble and enough water is present, the attractions between ions is not great enough to cause the ions to attach and form a crystalline solid. The ions remain in solution.

When we mix solutions containing dissolved ionic compounds, all the ions will intermingle and mutually interact. If any of the ions that are mixed constitute the ions of an insoluble ionic compound, the ions will combine to form crystals and the insoluble solid will separate from the solution. A reaction of ions in a solution to form an insoluble solid is called an ion combination or precipitation reaction. The insoluble solid that separates from the solution is known as a precipitate.

A good example of a precipitation reaction is the reaction between calcium ions and carbonate ions.

$$Ca^{2+}(aq) \ + \ CO_3{}^{2-}(aq) \ \longrightarrow \ CaCO_3(s)$$

Solid calcium carbonate is formed from a solution containing high enough concentrations of the two ions. Hard water is natural water that contains relatively high concentrations of calcium ion and other ions. When you use hard water calcium carbonate can precipitate to form scale in hot water pipes and tea kettles. Some natural limestone deposits are precipitated calcium carbonate and the stalactites and stalagmites in limestone caves are calcium carbonate deposits.

In nature precipitation typically occurs quite slowly but in laboratory tests, where the concentrations of ions are relatively high, precipitation generally occurs rapidly. The precipitate forms the instant the solutions are mixed, and slowly settles from the solution. Precipitation usually produces numerous small crystals which take the form of a finely divided solid. Precipitates have a variety of colors and appearances. Precipitates may be flocculent (fluffy), gelatinous (jellylike), coagulated (curdy or lumpy), or crystalline (powdery or silky) in appearance.

How can a precipitation or ion combination reaction be predicted when solutions are mixed? This is done by knowing which ionic compounds are soluble and which are insoluble. As you know compounds vary in their solubilities. An ionic compound is considered to be soluble if enough of it can be dissolved in water to prepare a solution that is about 0.1 M or greater. An ionic compound that is not soluble enough to prepare a water solution that is 0.1 M is said to be insoluble. When the constituent ions of an insoluble compound are mixed in relatively high concentrations, precipitation occurs.

The solubilities of many common ionic compounds in water can be summarized in the form of a solubility list. Recall that typical ionic compounds include metal ions in combination with nonmetal or polyatomic anions. For our purposes, the solubility list on the next page will be useful. Keep in mind that this list does not include all ionic compounds and that there are some exceptions to the general solubilities in the list. Refer to this list or to Table 20-1 when you want to check the solubility of an ionic compound.

1. Almost all ionic compounds containing sodium ion, Na^+, potassium ion, K^+, or ammonium ion, NH_4^+, are soluble.

2. Almost all ionic compounds containing nitrate ion, NO_3^-, or acetate ion, $C_2H_3O_2^-$, are soluble.

3. All common ionic compounds containing sulfate ion, SO_4^{2-}, are soluble except

> calcium sulfate, $CaSO_4$ strontium sulfate, $SrSO_4$
> barium sulfate, $BaSO_4$ and   lead(II) sulfate, $PbSO_4$

4. All common ionic compounds containing chloride ion, Cl^-, bromide ion, Br^-, iodide ion, I^-, and fluoride ion, F^-, are soluble except

silver, mercury(I), lead(II) chloride: $AgCl$, Hg_2Cl_2, $PbCl_2$

silver, mercury(I), lead(II) bromide: $AgBr$, Hg_2Br_2, $PbBr_2$

silver, mercury(I), lead(II) iodide: AgI, Hg_2I_2, PbI_2

magnesium, calcium, strontium, barium, lead(II) fluoride:
MgF_2, CaF_2, SrF_2, BaF_2, PbF_2

5. All common ionic compounds containing hydroxide ion, OH^-, are insoluble except

sodium hydroxide, $NaOH$; potassium hydroxide, KOH; barium hydroxide, $Ba(OH)_2$

6. Most ionic compounds containing carbonate ion, CO_3^{2-}, chromate ion, CrO_4^{2-}, and phosphate ion, PO_4^{3-}, are insoluble except those containing sodium ion, potassium ion (and other Group IA metal ions), or ammonium ion.

Table 26-1

Solubility List

Anion	Cations That Form Precipitates	Cations That Do Not Form Precipitates
NO_3^-	None	All other common
$C_2H_3O_2^-$	None	All other common
SO_4^{2-}	Ca^{2+}, Sr^{2+}, Ba^{2+}, Pb^{2+}	Most other common
Cl^-, Br^-, I^-	Ag^+, Hg_2^{2+}, Pb^{2+}	Most other common
OH^-	Most	Na^+, K^+, Ba^{2+}
F^-	Mg^{2+}, Ca^{2+}, Sr^{2+}, Ba^{2+}, Pb^{2+}	Most other common
CO_3^{2-}, CrO_4^{2-}, PO_4^{3-}	Most	Na^+, K^+, NH_4^+

Almost all ionic compounds of Na^+, K^+ and NH_4^+ are soluble

It is possible to predict whether a precipitation reaction will occur by considering which ions are involved and deciding whether any cations and anions will combine to form an insoluble compound. Of course, we can use the solubility list to decide which compounds are insoluble. As long as the concentrations of ions are high enough, insoluble compounds will be precipitated from the solution when the ions are mixed. Remember, whenever we are dealing with a solution of an ionic compound, we represent the compound in solution by writing the cation and anion as separate species.

Here are some examples with comments.

Write the balanced equation for any reaction that occurs when a solution of silver nitrate, $AgNO_3$, is added to a solution of calcium chloride, $CaCl_2$.

Since both solutions involve ionic compounds, we first write the formulas of the ions. Each cation and anion is written separately.

$$Ag^+ \ + \ NO_3^- \ + \ Ca^{2+} \ + \ 2Cl^-$$

Now we decide whether any of the cations will react with any anions to form an insoluble compound. According to solubility rule 2, all nitrates are soluble, so calcium nitrate will not precipitate. However, according to solubility rule 4, silver chloride is insoluble. Therefore, we predict the combination of the silver ions and chloride ions to form a silver chloride precipitate. The other ions are spectator ions so they will not react but remain in solution. The balanced equation for the reaction that occurs is

$$Ag^+ + \ + \ Cl^- \longrightarrow AgCl(s)$$

Since each ion carries a single charge, one silver ion reacts with one chloride ion to give silver chloride. In an equation involving ions, it is important to balance the charge as well as the number of atoms of each element.

Write the balanced equation for any reaction that occurs when a solution of aluminum nitrate, $Al(NO_3)_3$, is added to a solution of sodium hydroxide, NaOH.

First, we write the formulas of the ions present in the solutions.

$$Al^{3+} \ + \ 3NO_3^- \ + \ Na^+ + \ + \ OH^-$$

Now we decide whether any cation and anion combinations can occur. Sodium ion and nitrate ion do not react, since sodium nitrate is soluble. According to solubility rule 5, aluminum hydroxide is insoluble. So, we predict the reaction

$$Al^{3+} \ + \ 3OH^- \longrightarrow Al(OH)_3(s)$$

Of course, to balance the equation, three OH^- are needed to react with each Al^{3+}.

Write the balanced equation for any reaction that occurs when a solution of potassium acetate, $KC_2H_3O_2$, is added to a solution of sodium chloride, NaCl.

The ions involved are the cations and anions present in the solutions of the two compounds. These ions are

$$K^+ \ + \ C_2H_3O_2^- \ + \ Na^+ \ + \ Cl^-$$

Now we consider any possible reactions. Sodium ion will not react with acetate ion according to rules 1 and 2. Potassium ion will not react with chloride ion according to rules 1 and 4. Thus, we predict no precipitation reactions in this case, and the mixing of the two solutions merely produces a solution consisting of all four ions.

Write the balanced equation for any reaction that occurs when a solution of barium hydroxide, $Ba(OH)_2$, is mixed with a solution of zinc sulfate, $ZnSO_4$.

The ions present in the solutions are: $Ba^{2+} \ + \ 2OH^- \ + \ Zn^{2+} \ + \ SO_4^{2-}$

According to rule 3, $BaSO_4$ is insoluble, and according to rule 5, $Zn(OH)_2$ is insoluble. In this case two ion combination reactions occur. The equation showing both reactions is

$$Ba^{2+} \ + \ 2OH^- \ + \ Zn^{2+} \ + \ SO_4^{2-} \ \longrightarrow \ BaSO_4(s) \ + \ Zn(OH)_2(s)$$

Or we could give a separate equation for each reaction.

$$Ba^{2+} \ + \ SO_4^{2-} \ \longrightarrow \ BaSO_4(s)$$

$$Zn^{2+} \ + \ 2OH^- \ \longrightarrow \ Zn(OH)_2(s)$$

Laboratory Procedure

1. SOME PRECIPITATION REACTIONS

For each part, mix the solutions indicated, record any evidence of a reaction, briefly describe the color and appearance of any precipitate, and write a balanced net-ionic equation for any reaction. Record your observations and equations. Carry out the mixing in small, clean test tubes and use a clean glass rod for stirring the mixtures. You can estimate solution volumes.

184

(a) Place about 1 mL of a 0.1 M sodium chloride solution into a test tube. Pour in about 1 mL of a 0.1 M silver nitrate solution. Hint: The ions that are mixed are $Na^+ + Cl^-$ and $Ag^+ + NO_3^-$.

(b) Place about 1 mL of a 0.1 M barium nitrate solution into a test tube. Pour in about 1 mL of a 0.1 M sodium sulfate solution.

(c) Place about 1 mL of a 0.1 M silver nitrate solution into a test tube. Pour in about 1 mL of a 0.1 M potassium iodide solution.

(d) Place about 1 mL of a 0.1 M iron(III) chloride solution in a test tube. Pour in about 2 mL of a 0.1 M sodium hydroxide solution.

(e) Place about 1 mL of a 0.1 M sodium chloride solution into a test tube. Pour in about 1 mL of a 0.1 M potassium nitrate solution.

(f) Place about 1 mL of a 0.1 M sodium carbonate solution into a test tube. Pour in about 1 mL of a 0.1 M calcium chloride solution.

(g) Place about 1 mL of a 0.1 M barium hydroxide solution into a test tube. Pour in about 1 mL of a 0.1 M copper(II) sulfate solution.

(h) Place about 1 mL of a 0.1 M barium hydroxide solution into a test tube. Pour in about 1 mL of a 0.1 M sodium phosphate solution.

2. FOUR UNLABELED BOTTLES (OPTIONAL)

Four bottles of solutions are provided without labels. They are marked A, B, C and D for reference. The possible contents of the bottles are solutions of $AgNO_3$, NaCl, KI, and $NaNO_3$. By mixing various combinations of the four solutions, the contents of the bottles can be identified by observing any precipitation reactions that occur upon mixing. Using the solubility list and your knowledge of the appearance of certain precipitates gained in part one of this exercise, you can deduce which bottle contains which solution.

To begin, consider the prediction of any reactions that will occur when any two of the above solutions are mixed. Give the balanced net-ionic equations for any reactions. Not all combinations will give precipitation reactions. The six possible mixtures are given on the next page.

(a) AgNO$_3$ solution and NaCl solution

(b) AgNO$_3$ solution and KI solution

(c) AgNO$_3$ solution and NaNO$_3$ solution

(d) NaCl solution and KI solution

(e) NaCl solution and NaNO$_3$ solution

(f) KI solution and NaNO$_3$ solution

To determine the contents of the bottles, use the following approach. Place six small, clean test tubes in a test tube rack. Label the tubes or the rack so that each tube can be distinguished as either AB, AC, AD, BC, BD or CD. Instead of tubes you can lay a flat piece of plastic wrap on your desk and mix the solutions on various parts of the plastic using 3 to 5 drops. Using the set of unlabeled bottles add the following:

Tube AB	10 drops of solution A	
Tube AC	10 drops of solution A	
Tube AD	10 drops of solution A	
Tube AB	10 drops of solution B	
Tube BC	10 drops of solution B	
Tube BD	10 drops of solution B	
Tube AC	10 drops of solution C	
Tube BC	10 drops of solution C	
Tube CD	10 drops of solution C	
Tube AD	10 drops of solution D	
Tube BD	10 drops of solution D	
Tube CD	10 drops of solution D	

Make sure that the contents of each tube are well mixed and describe the appearance and color of any precipitates that form. Record your observations in the table including the observation of no observable reaction. Using your results, identify the contents of the four bottles. Explain the reasons for your choices.

A _____   B_____   C_____   D _____

Name _____

Lab Section _____ Due Date _____

1. Some Precipitation Reactions (observations and equations)

     a.

     b.

     c.

     d.

     e.

     f.

     g.

     h.

2. Four Unlabeled Bottles

    (a) $AgNO_3$ solution and NaCl solution

    (b) $AgNO_3$ solution and KI solution

    (c) $AgNO_3$ solution and $NaNO_3$ solution

    (d) NaCl solution and KI solution

    (e) NaCl solution and $NaNO_3$ solution

    (f) KI solution and $NaNO_3$ solution

A _____   B _____

C _____   D _____

26 QUESTIONS EXPERIMENT 26

Name _____

Lab Section _____ Due Date _____

Give balanced net-ionic equations for any predicted reactions when the following solutions are mixed. Use a solubility list. (NOTE: Solutions containing mercury and lead are poisonous and should never be poured down the drain.)

(a) A $Hg_2(NO_3)_2$ solution is added to a NaCl solution.

(b) A NH_4Br solution is added to a $AgNO_3$ solution.

(c) A Na_2CO_3 solution is added to a $CaCl_2$ solution.

(d) A $Pb(NO_3)_2$ solution is added to a NaBr solution.

(e) A $BaCl_2$ solution is added to a KF solution.

(f) A KI solution is added to a $AgNO_3$ solution.

(g) A NaOH solution is added to a $BaCl_2$ solution.

(h) A Na_2SO_4 solution is added to a $Pb(NO_3)_2$ solution.

(i) A $MgSO_4$ solution is added to a $Ca(OH)_2$ solution.

27 Acids and Bases

Objective

The purpose of this experiment is to observe the properties and reactions of the aqueous solution of some common acids and bases.

Terms to Know

Acid - A species that can lose a proton in a chemical reaction. A proton donor.

Base - A species that can gain a proton in a chemical reaction. A proton acceptor.

Conjugate Acid-Base Pair - An acid and base that are related by the loss and gain of a proton.

Strong Acid - An acid that reacts extensively with water to form hydronium ion and the conjugate base of the acid.

Weak Acid - An acid that reacts with water to a slight extent forming a solution with a relatively low concentration of hydronium ion.

Acid-Base Reaction - A chemical reaction in which an acid reacts with a base to give the conjugate base of the acid and the conjugate acid of the base. A proton transfer reaction.

Acid-Base Indicator - A chemical that turns a specific color in an acidic solution and a different color in a basic solution.

Acid-Base Table - A table which lists acids according to decreasing strengths and the conjugate bases of the acids according to increasing strengths.

Discussion

Acids and bases are very common chemicals and we use various acid solutions and base solutions as laboratory chemicals. According to the Brønsted-Lowry theory of acids and bases, an acid is a species that has a tendency to lose a proton (H^+) in a chemical reaction. An acid is a proton donor. A base is a species that has a tendency to gain a proton in a chemical reaction. A base is a proton acceptor. These definitions can be illustrated by a general equation.

$$H\text{-}A + B \longrightarrow A^- + H\text{-}B^+$$

in which the acid, HA, loses a proton to the base, B. An actual example is represented by the equation

$$HCl(g) + H_2O \longrightarrow H_3O^+(aq) + Cl^-(aq)$$

in which the acid hydrogen chloride loses a proton to the base water to form hydronium ion and chloride ion. When an acid reacts with a base and a proton transfer occurs, the reaction is called an acid-base reaction. An acid loses a proton to a base which is capable of bonding to the proton more strongly than the original acid. Thus, we say that an acid, HA, can lose a proton to leave a weaker base, A^-, and a base, B, can gain a proton to form a weaker acid, HB^+. Each acid has a corresponding base and each base will have a corresponding acid. An acid and base that are related by the loss and gain of a proton are called a conjugate acid-base pair. The acid-base pairs are labeled in the following equation.

$$HCl(g) + H_2O \longrightarrow H_3O^+(aq) + Cl^-(aq)$$

Acid 1   Base 2   Acid 2   Base 1
\   \ conjugate pair /   /
\   conjugate pair   /

An acid-base reaction will always involve an acid losing a proton to a base to form the conjugate base of the acid and the conjugate acid of the base. A proton is transferred in the process.

Many common chemicals act as acids or bases. Ammonia dissolved in water to form aqueous ammonia, $NH_3(aq)$, is a base. The soluble ionic compounds sodium hydroxide, NaOH, potassium hydroxide, KOH, and barium hydroxide, $Ba(OH)_2$, dissolve to form hydroxide ion, $OH^-(aq)$, and the corresponding metal ion. The hydroxide ion is a base. The hydrogen-containing compounds HCl, H_2SO_4, HNO_3, H_2S, $H_2C_2O_4$, $HC_2H_3O_2$, and H_3PO_4 are a few examples of common acids.

Some acids are called strong acids since, when they dissolve in water, they react completely with water to form hydronium ion and the conjugate base of the acid. In general, a strong acid, H-X, will react in water as

$$H\text{-}X + H_2O \longrightarrow H_3O^+(aq) + X^-(aq)$$

Thus, the aqueous solutions of the strong acids will always contain hydronium ion and the conjugate base of the acid. Strong acids are strong electrolytes since they form solutions of ions in water. The common strong acids are HCl, H_2SO_4 and HNO_3. Other acids react only slightly with water when dissolved in water:

$$\text{H-W} + H_2O \rightleftharpoons H_3O^+(aq) + W^-(aq)$$

Acids that react only slightly with water are called weak acids. Molecular weak acids are weak electrolytes. Solutions of weak acids will contain relatively low concentrations of hydronium ion and the conjugate base of the acid, and the major species in solution will be the acid. Thus, solutions of weak acids are represented by the formula of the acid. The species present in some common acid and base solutions are shown in Table 27-1. Use this table for reference.

Table 27-1

Some Acid and Base Solutions

Strong Acids

sulfuric acid	$H_3O^+ + HSO_4^-$
nitric acid	$H_3O^+ + NO_3^-$
hydrochloric acid	$H_3O^+ + Cl^-$

Weak Acids

phosphoric acid	$H_3PO_4(aq)$	
carbon dioxide	$CO_2(aq)$	(Use $CO_2 + H_2O$ for acid-base reactions)
acetic acid	$CH_3COOH(aq)$ or $HC_2H_3O_2(aq)$	
dihydrogen phosphate ion	$H_2PO_4^-(aq)$	(Prepared by dissolving NaH_2PO_4 or KH_2PO_4 in water.)
ammonium ion	$NH_4^+(aq)$	(Prepared by dissolving NH_4Cl in water.)

Bases

benzoate ion	$C_6H_5COO^-(aq)$	(Prepared by dissolving a soluble compound like NaC_6H_5COO in water.)
ammonia	$NH_3(aq)$	
carbonate ion	$CO_3^{2-}(aq)$	(Prepared by dissolving a soluble compound like Na_2CO_3 in water.)
hydroxide ion	$OH^-(aq)$	(Prepared by dissolving NaOH or KOH in water.)

When a solution of an acid is mixed with a solution of a base, a reaction will occur if a weaker acid and base can be formed. Table 27-2 is an acid-base table that lists some common acids and bases according to relative strengths. The acids in the table are listed in order of decreasing strength with the strongest acids at the top. Note that the three common strong acids are listed at the top of the table. Bases in the table are listed in order of increasing strength. The stronger bases are at the bottom of the table. We can use the table to predict acid-base reactions using the principle that any acid will react with a base that is below it in the table to give the conjugate base of the acid and the conjugate acid of the base.

Table 27-2

Acid-Base Table

	Acids		Bases
Sulfuric acid	H_2SO_4	HSO_4^-	Hydrogen sulfate ion
Hydrogen chloride	HCl	Cl^-	Chloride ion
Nitric acid	HNO_3	NO_3^-	Nitrate ion
Hydronium ion	H_3O^+	H_2O	Water
Oxalic acid	$H_2C_2O_4$	$HC_2O_4^-$	Hydrogen oxalate ion
Hydrogen sulfate ion	HSO_4^-	SO_4^{2-}	Sulfate ion
Phosphoric acid	H_3PO_4	$H_2PO_4^-$	Dihydrogen phosphate ion
Hydrogen fluoride	HF	F^-	Fluoride ion
Hydrogen oxalate ion	$HC_2O_4^-$	$C_2O_4^{2-}$	Oxalate ion
Acetic acid	$HC_2H_3O_2$	$C_2H_3O_2^-$	Acetate ion
Carbon dioxide (aq)	$(CO_2 + H_2O)$	HCO_3^-	Hydrogen carbonate ion
Hydrogen sulfide	H_2S	HS^-	Hydrogen sulfide ion
Dihydrogen phosphate ion	$H_2PO_4^-$	HPO_4^{2-}	Hydrogen phosphate ion
Hydrogen sulfite ion	HSO_3^-	SO_3^{2-}	Sulfite ion
Ammonium ion	NH_4^+	NH_3	Ammonia
Hydrogen cyanide	HCN	CN^-	Cyanide ion
Hydrogen carbonate ion	HCO_3^-	CO_3^{2-}	Carbonate ion
Hydrogen phosphate ion	HPO_4^{2-}	PO_4^{3-}	Phosphate ion
Hydrogen sulfide ion	HS^-	S^{2-}	Sulfide ion
Water	H_2O	OH^-	Hydroxide ion

Acids are listed according to decreasing strength. Bases are listed according to increasing strength.

To write a net-ionic equation representing an acid-base reaction, we write the acid and base that react on one side and the conjugate acid and base on the other. Table 27-2 can be used to predict the reaction. For example, consider the acid-base reaction that occurs when a solution of sodium hydroxide is mixed with a solution of the weak acid hydrogen fluoride, HF. Hydrogen fluoride solutions are toxic and dangerous. CAUTION: DO NOT WORK WITH HYDROGEN FLUORIDE SOLUTIONS IN THE LABORATORY. First, write the formulas of the species present in the solutions. The solution of the ionic compound sodium hydroxide contains the separate ions: $Na^+(aq)$ + $OH^-(aq)$. The hydrogen fluoride solution contains HF(aq) since HF is a weak acid. (Note the position of HF in Table 27-2.) The species in the solutions are:

$$Na^+(aq) + OH^-(aq) + HF(aq)$$

Finding HF and OH^- in Table 27-2, we note that the acid is above the base so we predict that they will react to give the weaker acid, H_2O, and the weaker base, F^-.

$$HF(aq) + OH^-(aq) \rightleftharpoons H_2O + F^-(aq)$$

As another example consider the acid base reaction that occurs when nitric acid and potassium hydroxide solutions are mixed. First, write the formulas of the species present in the solutions which are mixed. Use Table 27-1 as a guide. The major species present in nitric acid and potassium hydroxide are:

$$H_3O^+(aq) + NO_3^-(aq) + K^+(aq) + OH^-(aq)$$

In Table 27-2 we note that H_3O^+ is an acid and OH^- is a base. The reaction that occurs is

$$H_3O^+(aq) + OH^-(aq) \rightleftharpoons 2H_2O$$

In fact, this is the reaction that occurs when any strong acid solution is mixed with a solution containing hydroxide ion. This type of reaction forms water and is called neutralization.

Indicators are chemicals that take on different colors depending upon the concentration of hydronium ion or hydroxide ion in an aqueous solution. Some indicators are used to indicate whether a solution is acidic or basic. The indicator called litmus turns blue in a solution of a base and red in a solution of an acid. The indicator phenolphthalein is colorless in acidic solutions and turns pink in basic solutions. This indicator is used to observe the occurrence of acid-base reactions by adding indicator to an acid solution and then adding base solution until a pink color is noted. The idea is that the appearance of the pink color indicates the change from an acidic solution to a basic solution.

Laboratory Procedure

1. ACIDIC AND BASIC SOLUTIONS

Using seven clean test tubes, obtain about 0.5 mL of each of the solutions listed in part 1 on the report sheet. Place the test tubes in your test tube rack. Obtain seven pieces of red litmus paper and seven pieces of blue litmus paper and place them on a paper towel. Use a glass stirring rod to transfer a drop of the hydrochloric acid solution to the end of one piece of red litmus and a drop to the end of one piece of blue litmus. Record your observations of the color change in the space provided on the report sheet. Wipe off the stirring rod and repeat using the other solutions. Be sure to wipe off the rod each time and to record the results.

To each of the various solutions, add a drop of phenolphthalein indicator. Record the color in each solution in the space provided on the report sheet.

2. ACID-BASE REACTIONS

For each of the following, note the formulas of the major species present in the solutions before mixing and give a balanced equation for any reaction. For your convenience, you do not have to include the parenthetical aq after each formula when you write equations. Refer to Tables 27-1 and 27-2 as needed. For efficiency, calibrate several small test tubes by pouring 2 mL of water in them and marking the 2 mL level with tape or waxed pencil. Rinse test tubes and corks with deionized water after each use. For litmus tests, use strips of litmus paper upon which drops of the solutions can be tested using a clean stirring rod. Record your observations and equations.

(a) Pour 2 mL of a 1 M hydrochloric acid solution into one clean test tube and slightly more than 2 mL of a 1 M sodium hydroxide solution in another. Put one drop of phenolphthalein indicator into the acid solution and note the color. Pour the base solution into the acid solution and note any evidence for a reaction. Color _____

The formulas for the major species in the solutions before mixing are

$$H_3O^+ + Cl^- \text{ and } Na^+ + OH^-.$$

Write a balanced net-ionic equation for the acid-base reaction that took place.

(b) Place 2 mL of a 1 M acetic acid solution in a clean test tube and test with litmus. Pour slightly more than 2 mL of a 1 M sodium hydroxide solution into the acetic acid. Cork, shake and test with litmus.

Litmus tests:

The species mixed are: $HC_2H_3O_2$ and $Na^+ + OH^-$

Balanced equation:

(c) Place 2 mL of a 1 M sodium acetate solution in a test tube and test with litmus. Add slightly more than 2 mL of a 1 M hydrochloric acid solution to the tube. Cork, shake and test with litmus.

Litmus tests:

The species mixed are: $Na^+ + C_2H_3O_2^-$ and $H_3O^+ + Cl^-$

Balanced equation:

(d) Place 2 mL of a 1 M ammonia solution in a test tube and test with litmus. Add slightly more than 2 mL of a 1 M hydrochloric acid solution to the tube. Cork, shake vigorously and test with litmus.

Litmus tests:

The species mixed are: NH_3 and $H_3O^+ + Cl^-$

Balanced equation:

(e) Place 2 mL of a 1 M ammonium chloride solution in a test tube and test with litmus. Add slightly more than 2 mL of a 1 M sodium hydroxide solution to the tube. Cork, shake vigorously and test with litmus. Avoid breathing the fumes.

Litmus tests:

The species mixed are: $NH_4^+ + Cl^-$ and $Na^+ + OH^-$

Balanced equation:

(f) Place a small sample of solid benzoic acid, HC_6H_5COO, about the size of a dried pea in a test tube. Add about 1 mL of deionized water, cork and shake vigorously. Comment on the solubility of the acid in water. Pour about 2 mL of a 1 M sodium hydroxide solution in the test tube, cork and shake vigorously for a few minutes. Describe what happens and test with litmus. Benzoic acid is not in the acid-base table but it is a weak acid.

Solubility in water:

 Litmus test:

Give a balanced equation to show what happened. (Hint: Benzoic acid is a weak acid not included in the acid-base table and benzoate ion is a product of the reaction. See Table 27-1.)

(g) Place 1 mL of a 1 M sodium benzoate (See Table 27-1) solution in a test tube and test with litmus. Add 2 mL of a 1 M hydrochloric acid solution to the tube, cork and shake. Describe the results and test with litmus. Benzoate ion is not in the acid-base table but it is a the conjugate base of benzoic acid (See part (f) above) .

Litmus test:

Result:

Give a balanced equation to show what happened.

(h) Pour 2 mL of a 1 M sodium carbonate solution into a small beaker and test with litmus. Carefully and slowly add about 5 mL of a 1 M hydrochloric acid solution and describe the results. (Hint: Two acid-base reactions occur. One product of the first reaction is involved in the second reaction and the bubbles are CO_2 gas formed in the second reaction.)

Litmus test:

Result:

The species mixed are:

Balanced equation: (Hint see Table 27-1 for the species in a solution of sodium carbonate.)

27 REPORT SHEET EXPERIMENT 27

Name _____

Lab Section _____ Due Date _____

1. Acidic and Basic Solutions.

Solution	Litmus	Phenolphthalein
1 M HCl, hydrochloric acid	_____	
1 M H_2SO_4, sulfuric acid	_____	
1 M NH_4^+, ammonium ion (from NH_4Cl)	_____	
1 M $HC_2H_3O_2$, acetic acid	_____	
1 M NaOH, sodium hydroxide	_____	
1 M NH_3, ammonia	_____	
1 M CO_3^{2-}, carbonate ion (from Na_2CO_3)	_____	

2. Acid-Base Reactions

(a) Color after mixing _____

   Write a balanced net-ionic equation for the acid-base reaction that took place.

(b) Litmus tests:

Balanced equation:

(c) Litmus tests:

Balanced equation:

(d) Litmus tests:

Balanced equation:

(e) Litmus tests:

Balanced equation:

(f) Solubility in water:

   Litmus test:

Give a balanced equation to show what happened:

(g) Litmus test:

Result:

Give a balanced equation to show what happened:

(h) Litmus test:

Result:

Give balanced equations to show what happened:

27 QUESTIONS EXPERIMENT 27

Name _____

Lab Section _____ Due Date _____

1. Give definitions for:

(a) Acid

(b) Base

2. Give balanced net-ionic equations for any acid-base reactions that occur upon mixing the following solutions. Refer to Tables 27-1 and 27-2.

(a) A solution of sodium hydrogen carbonate, $NaHCO_3$, is mixed with a solution of sodium hydroxide, NaOH.

(b) A solution of sodium dihydrogen phosphate, NaH_2PO_4, is mixed with a solution of nitric acid.

(c) A solution of ammonia, NH_3, is mixed with a solution of sodium hydroxide, NaOH.

(d) A solution of potassium hydrogen oxalate, KHC_2O_4, is mixed with sulfuric acid.

28 Acid-Base Titration

Objective

To practice the titration method of analysis and to use the method to analyze a vinegar solution and a hydrochloric acid solution.

Terms to Know

Titration - The process of adding a measured amount of a solution of known concentration to a sample of another solution for purposes of determining the concentration of the solution or the amount of some species in the solution.

Titrant - The solution of known concentration used in a titration.

Buret - A narrow, cylindrical-shaped, precisely calibrated piece of glassware. A device used to measure the volume of titrant delivered in a titration.

End Point - The point in a titration when just enough titrant has been added to react with all of the titrated species in the solution being titrated.

Indicator - A chemical, added to a titration mixture, which changes color at the end point of the titration.

Discussion

A titration is the process of adding a measured volume of a solution of known concentration to a sample of another solution for purposes of determining the concentration of the second solution or the amount of some species in the solution. A species in the solution of known concentration reacts with another species in the unknown solution. The addition and measurement of the volume of the solution of known concentration is carried out by use of a buret. (See Fig. 28-1.) A titration is usually carried out by placing a measured sample of the unknown solution in a flask, filling the buret with the known solution (called the titrant) and then slowly delivering the titrant to the flask until the necessary amount has been added to the unknown solution.

The point at which the necessary amount has been added is called the end point of the titration. The end point is often detected by placing a small amount of a chemical called an indicator in the reaction flask. The indicator is chosen so that it will react with the titrant when the end point is reached. The reaction of the indicator produces a colored product; the appearance of the color signals the end point of the titration. Some indicators are colored to begin with and react at the end point to produce a different colored product, so the change in color indicates the end point. Other indicators change from a colorless form to a colored form at the end point.

Once the end point has been found, the volume of titrant used can be determined from the buret. Using this volume, the concentration of the titrant and the stoichiometric factor from the balanced equation, we can deduce the number of moles of species in the solution being titrated. If the molarity of the unknown solution is to be calculated, it is necessary to measure the volume of the original sample of unknown solution before it was titrated. Then, the molarity can be found by dividing calculated number of moles by this volume. As an example, consider the following case. Determine the molarity of a hydrochloric acid solution if 30.21 mL of a 0.200 M sodium hydroxide solution is needed to titrate a 25.00 mL sample of the acid solution. First, the chemical reaction involved is

$$OH^-(aq) + H_3O^+(aq) \rightleftharpoons 2H_2O$$

Note that one mole of acid reacts with one mole of base. The number of moles of hydroxide ion needed to react can be found from the volume used and the molarity. The titration required 30.21 mL of 0.200 M NaOH. The number of moles of hydroxide ion is found by multiplying the volume of sodium hydroxide solution in liters by the molarity. The calculations are:

First the number of moles of OH^- used are found as the product of the volume and molarity: $V_b M_b$

$$30.21 \text{ mL} \quad \frac{0.200 \text{ mol } OH^-}{1000 \text{ mL}} \quad = \quad 0.006042 \text{ mol } OH^-$$

Next the number of moles of hydronium ion is found by multiplying by the molar ratio obtained from the equation for the titration reaction.

$$0.006042 \text{ mol } OH^- \quad \frac{1 \text{ mol } H_3O^+}{1 \text{ mol } OH^-} \quad = \quad 0.006042 \text{ mol } H_3O^+$$

Finally the molarity of the acid solution can be found by dividing the number of moles of hydronium ion by the original volume of the acid solution in liters.

$$\frac{0.006042 \text{ mol } H_3O^+}{25.00 \text{ mL}} \quad \frac{1000 \text{ mL}}{1 \text{ L}} \quad = \quad 0.242 \text{ M } H_3O^+$$

If the acid and base in a titration react in a one-to-one molar ratio, as is the case in the above example, the calculations can be simplified by using the equation:

$$M_a = \frac{M_b V_b}{V_a}$$

where V_b is the volume of the base used in the titration, V_a is the volume of the original acid solution, M_b is the molarity of the base solution and M_a is the molarity of the acid solution. Using this simple equation for the example given above the molarity of the acid solution is:

$$M_a = 0.200 \text{ M} \frac{30.21 \text{ mL}}{25.00 \text{ mL}} = 0.242 \text{ M}$$

HOW TO USE THE BURET

A buret is a piece of glass tubing calibrated to deliver measured volumes of solution. A 50-mL buret is calibrated to deliver between 0 and 50 mL. Each etched line on the buret corresponds to a 0.1 mL increment of volume. By interpolation, the buret can be read to the nearest 0.01 mL. These instructions apply to a Mohr buret as pictured in Figure 28-1. If you use a buret with a glass or plastic stopcock, disregard the reference to the rubber tubing and bead but follow the other instructions.

1. CLEANING

Place some water in the buret and allow it to run out through the tip by squeezing the rubber tube around the glass bead (or opening the stopcock). If you notice water drops adhering to the sides of the buret as it is drained, the buret needs cleaning. Clean the buret using tap water and a small amount of detergent. Use a buret brush to scrub the inside of the buret. Rinse the buret four or five times with tap water and allow some water to run out of the tip each time. Finally, rinse the buret three or four times with 10 mL portions of distilled water. By rotating the buret, allow the water to rinse the entire buret and be sure to rinse the tip by allowing some water to pass through.

2. FILLING

When filling a buret, try not to splash or spill any titrant. Clean up any spills. Use a paper towel to dry the outside of the buret. Rinse the clean buret two or three times with less than 5 mL portions of the solution with which it is to be filled. Be sure to rinse the tip each time. Fill the buret to a level just above the 0.00 mL mark. Fill the tip by bending the rubber to point the tip upward and then gently squeeze the tube to allow the liquid to displace the air (or open the stopcock and remove any air bubbles). Adjust the meniscus to a position somewhat below the 0.00 mL mark.

3. READING

The position of the bottom of the meniscus is read to the nearest 0.01 mL. The buret can be read directly to the nearest 0.1 mL, but you must interpolate to read to the nearest 0.01 mL. To interpolate, you imagine that the distance between lines is made up of 10 equally spaced parts. Then decide with which part the bottom of the meniscus coincides. A buret reading card can be used to aid you in reading the position of the meniscus. (See Fig. 28-2.) When reading the buret, make sure your eye is level with the bottom of the meniscus.

4. TITRATION METHOD

Be sure to record the initial and final buret readings when you do a titration. Place the sample of solution to be titrated in a flask. After recording the initial volume of the buret, allow the titrant to flow into the flask by pinching the rubber tube around the glass bead (or opening the stopcock). Let the titrant flow rapidly at first, and then add smaller and smaller volume increments as the end point is approached. When you are close to the end point, add one drop or less at a time. A fraction of a drop can be added by allowing a portion of a drop to form on the tip, touching the tip to the inside of the flask and then washing down the sides of the flask with a small amount of distilled water from a wash bottle. During the titration, mix the solutions in the flask by swirling but do not splash the solutions out of the flask. (See Fig. 28-1b.)

5. CLEANING UP

Drain the solution from the buret and rinse thoroughly with tap water. Remember to rinse the tip. Try not to spill any of the solutions on the desk or your clothing. If you do spill the solutions, clean them up. If any solutions spill on your clothing, tell your instructor. Burets can be stored corked and filled with distilled water.

Laboratory Procedure

1. ANALYSIS OF ACETIC ACID IN VINEGAR

In this experiment you will titrate a vinegar solution using a standardized sodium hydroxide solution. You will need a clean, dry 100-mL beaker, a 125-mL flask, a 250-mL flask, a buret, a buret clamp and a ring stand. Set up a buret on a ring stand as shown in Fig. 28-1 but use only one buret. Obtain about 100-mL of a standardized sodium hydroxide solution in a 100-mL beaker. (DANGER: NaOH solutions are very caustic.) Record the molarity of this solution. Be very careful not to splash or spill this solution. Sodium hydroxide solutions are especially dangerous if splashed in your eyes. Always pour such solutions in a buret by removing the buret from the clamp and holding the buret in a piece of paper towel well below your eye level. Never raise the beaker of solution above your eye level. Rinse and fill your buret with the sodium hydroxide solution according to the instructions given in the discussion section.

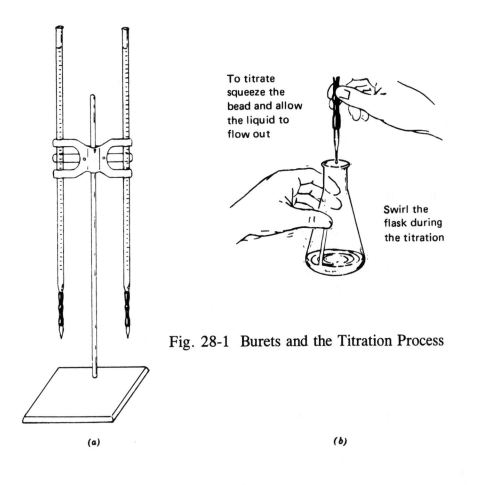

To titrate
squeeze the
bead and allow
the liquid to
flow out

Swirl the
flask during
the titration

(a)   (b)

Fig. 28-1 Burets and the Titration Process

Fig 28-2 Reading a buret.
(a) Keep eye level with the
   bottom of the meniscus.

(b) Use a buret reading
   card behind the buret
   to locate the bottom of
   the meniscus.

Enlarged buret
reads 32.54 ml.

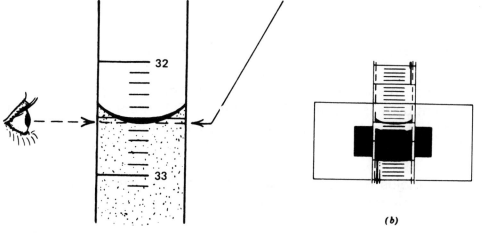

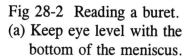

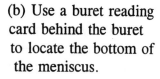

32

33

(a)

(b)

You will find vinegar solution in a plastic bottle fitted with a 10.00 mL pipet. As demonstrated by your instructor, pipet a 10 mL sample of vinegar into a clean 250 mL flask. Add 2 or 3 drops of phenolphthalein indicator to the flask and dilute to about 50 mL volume using distilled water. Read the initial volume of the buret. Titrate the acid solution sample with the sodium hydroxide solution until the end point is reached. As you add base, you will notice a slight pink color at the point at which the base solution enters the acid solution. As you approach the end point, you will see the entire solution flash pink. When this occurs, slow down the rate of addition of the base and carefully approach the end point. When the solution in the flask turns a very light pink and remains pink for one minute or longer you have reached the end point. The sides of the flask may be washed down with small amounts of distilled water at any time. Once the end point has been reached, read the final volume.

Rinse the 250-mL flask with distilled water. Pipet another 10 mL of acid into the flask and titrate the acid with the base. You may have to refill the buret with some sodium hydroxide solution. Record the titration data in the table on the Report Sheet. Rinse the flask and carry out a third titration. When you are finished be sure to rinse out your buret with water and remember to rinse the tip of the buret so that it does not contain base solution. Test the tip using red litmus paper. If the paper turns blue rinse your buret and tip with more water.

Use the volumes of the acid and base solutions used from the data table for the calculations. The acid-base reaction involved is

$$HC_2H_3O_2(aq) + OH^-(aq) \rightleftharpoons H_2O + C_2H_3O_2^-(aq)$$

Calculate the molarity of acetic acid in vinegar for each titration. If your results are within 10% of one another, calculate the average molarity. If your results do not seem close enough, consult your instructor.

2. THE PERCENT BY MASS ACETIC ACID IN VINEGAR

Assuming that the density of vinegar is 1.000 g/mL vinegar, it is possible to calculate the percent by mass acetic acid in vinegar. This is done by carrying out the following sequence of calculations.(AA is an abbreviation for acetic acid, $HC_2H_3O_2$)

$$\frac{\text{moles AA}}{\text{1 L vinegar}} \quad \text{to} \quad \frac{\text{moles AA}}{\text{1 mL vinegar}} \quad \text{to} \quad \frac{\text{moles AA}}{\text{1 g vinegar}} \quad \text{to} \quad \frac{\text{g AA}}{\text{1 g vinegar}} \times 100 = \%AA$$

Using your experimental value for the molarity of acetic acid in vinegar, calculate the percent by mass acetic acid in vinegar.

3. ANALYSIS OF AN UNKNOWN HYDROCHLORIC ACID SOLUTION

In this experiment you will titrate an unknown acid solution using a standardized sodium hydroxide solution. You will need a clean, dry 100-mL beaker, a 125-mL flask, a 250-mL flask, a 20-mL pipet, a pipet bulb, a buret, a buret clamp and a ring stand.

Set up a buret on a ring stand as shown in Fig. 28-1 but use only one buret. Obtain about 100-mL of a standardized sodium hydroxide solution in a 100-mL beaker. (DANGER: NaOH solutions are very caustic.) Record the molarity of this solution. Be very careful not to splash or spill this solution. Sodium hydroxide solutions are especially dangerous if splashed in your eyes. Always pour such solutions in a buret by removing the buret from the clamp and holding the buret in a piece of paper towel well below your eye level. Never raise the beaker of solution above your eye level. Rinse and fill your buret with the sodium hydroxide solution according to the instructions given in the discussion section.

Obtain a sample of unknown hydrochloric acid solution and record the number of the unknown. Using the pipet and pipet bulb as demonstrated by your instructor, measure 20.00 mL of unknown solution and place it in a 250-mL flask. Add 2 or 3 drops of phenolphthalein indicator to the flask and dilute to about 50 mL volume using distilled water. Read the initial volume of the buret. Titrate the acid solution sample with the sodium hydroxide solution until the end point is reached. As you add base, you will notice a slight pink color at the point at which the base solution enters the acid solution. As you approach the end point, you will see the entire solution flash pink. When this occurs, slow down the rate of addition of the base and carefully approach the end point. When the solution in the flask turns a very light pink and remains pink for one minute or longer you have reached the end point. The sides of the flask may be washed down with small amounts of distilled water at any time. Once the end point has been reached, read the final volume.

Rinse the 250-mL flask with distilled water. Pipet another 20 mL of acid into the flask and titrate the acid with the base. You may have to refill the buret with some sodium hydroxide solution. Record the titration data in the table on the Report Sheet. Rinse the flask and carry out a third titration. When you are finished be sure to rinse out your buret with water and remember to rinse the tip of the buret so that it does not contain base solution. Test the tip using red litmus paper. If the paper turns blue rinse your buret and tip with more water.

Use your titration data to calculate the molarity of the unknown hydrochloric acid solution. If your results are within about 10% of one another, calculate the average molarity. If your results do not seem close enough, consult your instructor.

28 REPORT SHEET EXPERIMENT 28

Name _____

Lab Section _____ Due Date _____

TITRATION ANALYSIS USING MOLARITY

1. Analysis of Acetic Acid in Vinegar

Data Table:

Molarity of NaOH solution	_____			
Final volume NaOH solution	_____	_____	_____	_____
Initial volume NaOH solution	_____	_____	_____	_____
Volume NaOH solution used	_____	_____	_____	_____
Volume vinegar	10.00 mL	10.00 mL	10.00 mL	10.00 mL
Calculated molarity of acetic acid in vinegar	_____	_____	_____	_____

Average M acetic acid in vinegar

Give the setups for the molarity calculations below.

2. The Percent by Mass Acetic Acid in Vinegar from Molarity

Give the setup and calculated result.

3. Analysis of an Unknown Hydrochloric Acid Solution

Data Table:

Molarity of NaOH solution _____   Unknown Number _____

Final volume NaOH solution _____ _____ _____ _____

Initial volume NaOH solution _____ _____ _____ _____

Volume NaOH solution used _____ _____ _____ _____

Volume acid solution used   20.00 mL 20.00 mL 20.00 mL 20.00 mL

Calculated molarity acid   _____ _____ _____ _____
 solution

Average M of acid solution

Give the setups for the molarity calculations below.

28 QUESTIONS   EXPERIMENT 28

Name _____

Lab Section _____   Due Date _____

1. Explain the function and purpose of an indicator in a titration.

2. If you are titrating an acid with a base, tell how each of the following factors will affect the calculated molarity of the acid. That is, would the calculated molarity be greater than it should be, less than it should be, or not affected.

(a) The end point is exceeded by adding too much base.

(b) The volume of acid is measured incorrectly so that it is smaller than the value used in the calculations.

(c) Distilled water is used to wash down the sides of the flask during the titration.

3. A sample of vinegar is titrated with a sodium hydroxide solution to find the molarity of acetic acid. If 18.82 mL of a 0.395 M NaOH solution is required to titrate a 15.0 mL vinegar solution, what is the molarity of acetic acid in the vinegar?

29 Oxidation-Reduction Reactions

Objective

The purpose of this experiment is to observe some oxidation-reduction reactions occurring in aqueous solutions and to write equations for these reactions.

Terms to Know

Redox Reaction - A chemical reaction involving the oxidation of some species and the reduction of another species. Also known as an electron transfer reaction involving the loss of electrons by some species and the gain of electrons by another species.

Oxidation - The loss of electrons; the increase in oxidation number of an element.

Reduction - The gain of electrons; the decrease in oxidation number of an element.

Oxidizing Agent - A chemical species that can oxidize another species.

Reducing Agent - A chemical species that can reduce another species.

Redox Table - A table in which common oxidizing agents are listed according to decreasing strength and the corresponding reducing agents are listed according to increasing strength.

Discussion

Oxidation-reduction or redox reactions are electron transfer reactions. Oxidation is the loss of electrons by some species and reduction is the gain of electrons by some other species. In a redox reaction, oxidation and reduction occur simultaneously. That is, in a redox reaction, one reactant is oxidized and another is reduced.

When a redox reaction occurs, the oxidation number of one element increases and the oxidation number of another element decreases. Oxidation is characterized by an increase in oxidation number and reduction is characterized by a decrease in oxidation number. An oxidizing agent is a species that can oxidize another species and a reducing agent is a species that can reduce another species. When a solution containing an oxidizing agent is added to a solution containing a reducing agent, a redox reaction may occur. However, not all oxidizing and reducing agents will react with one another. A redox reaction is represented by a net-ionic equation showing the oxidizing agent, the reducing agent and the products they form along with any other species needed to balance the equation.

Equations for redox reactions can be predicted using redox tables such as those given in Table 29-1 and 29-2. To predict a reaction we use the principle that an oxidizing agent will react with any reducing agent that is below it in the redox table. To illustrate, consider the reaction that takes place when a solution of potassium permanganate (K^+ + MnO_4^-) in dilute sulfuric acid is mixed with a solution of sodium bromide (Na^+ + Br^-). The acid serves as a source of H_3O^+. To make balancing of equations easier the hydrogen ion, H^+, is used to represent H_3O^+. From Table 29-1 we can see that it is expected that an acidified solution of MnO_4^- will react as an oxidizing agent and oxidize the reducing agent, Br^-. The products of the reaction are manganese(II) ion, Mn^{2+}, and bromine, Br_2. To write the equation for the reaction, first write the MnO_4^- half reaction as it is found in the table and, below this, write the Br^- reaction in the reverse of that which is in the table.

$$MnO_4^- + 8H^+ + 5e^- \longrightarrow 4H_2O + Mn^{2+}$$

$$2Br^- \longrightarrow Br_2 + 2e^-$$

The total number of electrons lost and gained must be equal. To balance the electrons lost with the electrons gained, multiply the MnO_4^- half reaction by 2 and the Br^- half reaction by 5. Then add the two half reactions for the complete equation

$$2MnO_4^- + 16H^+ + 10e^- \longrightarrow 8H_2O + 2Mn^{2+}$$
$$10Br^- \longrightarrow 5Br_2 + 10e^-$$

$$2MnO_4^- + 16H^+ + 10Br^- \longrightarrow 5Br_2 + 8H_2O + 2Mn^{2+}$$

How do we know when a reaction has occurred in a solution? To "see" a reaction, some observable and noticeable change must accompany the reaction. Consider the following examples given on the next page that illustrate typical changes that can accompany redox reactions.

1. The consumption or dissolving of a solid:

$$Zn(s) + 2H^+ \longrightarrow Zn^{2+} + H_2$$

Solid zinc metal dissolves in acid.

2. The formation of a gas:

$$H_2O_2 + 2H^+ + 2Cl^- \longrightarrow Cl_2(g) + 2H_2O$$

Green-yellow chlorine gas is produced. (DANGER: Do not try this reaction.)

3. The formation of a solid:

$$2Al + 3Cu^{2+} \longrightarrow 3Cu(s) + 2Al^{3+}$$

Solid copper is formed as solid aluminum dissolves.

4. A color change (a colored reactant forms a colored product, a colored reactant is consumed or a colored product is formed):

$$8H^+ + Cr_2O_7^{2-} + 3H_2C_2O_4 \longrightarrow 2Cr^{3+} + 6CO_2 + 7H_2O$$

Yellow-orange $Cr_2O_7^{2-}$ changes to green Cr^{3+}.

Of course reactions other than redox can involve physical changes. Such changes are indications of chemical reactions. However, many redox reactions will display one or more of the changes describe here and we can use the changes as an indication that a reaction has occurred.

Laboratory Procedure

For the following exercises record your observations and give balanced equations.

1. OBSERVING REDOX REACTIONS

In this section, a redox reaction will be observed. Describe any evidence of a reaction and write an equation for the reaction.

Place a small piece of aluminum foil in a small test tube. Add about 2 mL of a 6 M hydrochloric acid solution to the tube. Allow the reaction to occur and observe any changes that you see taking place.

Table 29-1   *A Redox Table Including Some Common Oxidizing and Reducing Agents*

Oxidizing Agents		Reducing Agents
$H_2O_2 + 2H^+ + 2e^-$	$\rightleftarrows$	$4H_2O$
$MnO_4^- + 8H^+ + 5e^-$	$\rightleftarrows$	$Mn^{2+} + 4H_2O$
$Cl_2(g) + 2e^-$	$\rightleftarrows$	$2Cl^-$
$Cr_2O_7^{2-} + 14H^+ + 6e^-$	$\rightleftarrows$	$2Cr^{3+} + 7H_2O$
$Br_2(\ell) + 2e^-$	$\rightleftarrows$	$2Br^-$
$NO_3^- + 4H^+ + 3e^-$	$\rightleftarrows$	$NO(g) + 2H_2O$
$Fe^{3+} + 1e^-$	$\rightleftarrows$	Fe^{2+}
$I_2(s) + 2e^-$	$\rightleftarrows$	$2I^-$
$2CO_2(g) + 2H^+ + 2e^-$	$\rightleftarrows$	$H_2C_2O_4$

Table 29-2 *A Redox Table Including Common Metals*

Oxidizing Agents		Reducing Agents	
$Ag^+ + 2e^-$	$\rightleftarrows$	$Ag(s)$	
$Hg^{2+} + 1e^-$	$\rightleftarrows$	$Hg(\ell)$	
$Fe^{3+} + 1e^-$	$\rightleftarrows$	Fe^{2+}	
$Cu^{2+} + 2e^-$	$\rightleftarrows$	$Cu(s)$	
$2H^+ + 2e^-$	$\rightleftarrows$	$H_2(g)$	
$Pb^{2+} + 2e^-$	$\rightleftarrows$	$Pb(s)$	- - - - - - - - - - - - - - - - - - - -
$Ni^{2+} + 2e^-$	$\rightleftarrows$	$Ni(s)$	These metals dissolve in acid solutions
$Co^{2+} + 2e^-$	$\rightleftarrows$	$Co(s)$	
$Cd^{2+} + 2e^-$	$\rightleftarrows$	$Cd(s)$	
$Fe^{2+} + 2e^-$	$\rightleftarrows$	$Fe(s)$	
$Zn^{2+} + 2e^-$	$\rightleftarrows$	$Zn(s)$	
$2H_2O + 2e^-$	$\rightleftarrows$	$H_2(g) + 2OH^-$	
$Al^{3+} + 3e^-$	$\rightleftarrows$	$Al(s)$	
$Mg^{2+} + 2e^-$	$\rightleftarrows$	$Mg(s)$	- -
$Na^+ + 1e^-$	$\rightleftarrows$	$Na(s)$	These metals react spontaneously in water
$Ca^{2+} + 2e^-$	$\rightleftarrows$	$Ca(s)$	
$K^+ + 1e^-$	$\rightleftarrows$	$K(s)$	

As observed, aluminum dissolves, a gas is produced, and the solution turns cloudy and then clears. Since we started with solid aluminum and added hydrochloric acid, the formulas of the species that were mixed are Al and $H^+ + Cl^-$.

To explain the reaction that occurred, refer to Table 29-2. Note that aluminum metal is below hydrogen ion (H^+) in the table. Aluminum will react with hydrogen ion. To write the equation, use the half reaction of H^+ from the table and the reverse of the aluminum half reaction from the table. The predicted reaction is the sum of these two half reactions. Multiply each half by the proper number to make the number of electrons lost equal the number gained. Write the balanced net-ionic equation for the reaction between aluminum and hydrochloric acid.

2. SOME TYPICAL OXIDATION-REDUCTION REACTIONS

For each of the following, mix the solutions in a small test tube and stir with a clean glass rod. The volumes used need only be approximate. Record any changes and evidence of any reaction. Do not hurry, but rather allow time for any slower reactions to take place. Using the formulas of the species that are mixed, the half reactions can be found in a redox table and a balanced net-ionic equation can be written for any reaction that occurred. Refer to any redox table for half reactions.

(a) Place 1 mL of a fresh 3 % hydrogen peroxide solution in a test tube and add 1 mL of 1 M sulfuric acid. Now add some solid iron(II) ammonium sulfate (a sample about the size of a pencil eraser). Stir and observe any changes. The species mixed are
    $H_2O_2 + H^+ + HSO_4^-$ and $Fe^{2+} + NH_4^+ + SO_4^{2-}$. (Hint: The iron(II) ion is involved in the reaction but the ammonium ion and sulfate ion are not involved.)

(b) Place 1 mL of a 0.1 M potassium iodide solution in a test tube. Now add about 1 mL of 0.1 M iron(III) nitrate solution. Stir and observe any changes. The species mixed are $K^+ + I^-$ and $Fe^{3+} + NO_3^-$.

(c) Place 1 mL of a fresh 3% hydrogen peroxide solution in a test tube and add 1 mL of 1 M sulfuric acid. Work in the fume hood. Now add 1 mL of 0.1 M potassium iodide solution. Stir and observe any evidence of a reaction. The species mixed are
    $H_2O_2 + H^+ + HSO_4^-$ and $K^+ + I^-$.

(d) Pour about 0.5 mL of a 0.1 M silver nitrate solution in a test tube. Obtain a small length of copper wire and place it in the test tube so that the end dips into the solution. Allow the wire to remain in the solution for a few minutes. Remove the wire and inspect the end of the wire. Return the length of wire to the stockroom or to the designated container. The species mixed are $Ag^+ + NO_3^-$ and Cu.

(e) Pour about 1 mL of a 0.1 M copper(II) nitrate solution into a test tube. Drop a small piece of solid zinc metal into the test tube. Describe any changes. The species mixed are Cu^{2+} + NO_3^- and Zn.

(f) Pour 1 mL of a 1 M hydrochloric acid solution into a test tube. Drop in a small piece of magnesium metal. Be careful of any spattering. The species mixed are H^+ + Cl^- and Mg.

(g) Obtain a piece of calcium metal and, if necessary, rub it on a piece of sandpaper to remove some oxide coating and expose the metal. Add the metal to 2 mL of distilled water in a large test tube. Be careful that it does not bubble over. The species mixed are Ca and H_2O. (Hint: This is a redox reaction and a precipitation reaction. One of the products of the reaction is solid, white calcium hydroxide, $Ca(OH)_2$.)

3. THE HALOGENS AND HALIDE IONS

The halogens include fluorine, chlorine, bromine, and iodine. The halide ions are fluoride ion, chloride ion, bromide ion, and iodide ion. Elemental halogens are different from the halide ions. The halogens in the elemental form occur in diatomic molecular form and the halide ions occur as components of ionic compounds. Give the formulas of the diatomic molecules and ions of the various halogens.

Element	Symbol	Molecular Formula	Halide Ion Formula
Fluorine	F		
Chlorine	Cl		
Bromine	Br		
Iodine	I		

The elemental halogens occur in various colors. Fluorine is a pale-yellow gas but it is so chemically reactive that fluorine samples are seldom used in the laboratory. Samples of the other halogens are on display in the laboratory along with samples of solutions of the halogens in hexane. Observe the samples and record the colors.

Element	Color of Element	Color of Hexane Solution
Chlorine		
Bromine		
Iodine		

Some typical ionic compounds containing the halide ions and water solutions of these compounds are on display in the laboratory. Observe the display and fill in the following table:

Formula of Compound	Color and Form of Compound	Color of Water Solution
_____		
_____		
_____		
_____		

What do you conclude about the color of the halide ions?

4. PREPARING HALOGENS

In this exercise, samples of the halogens chlorine, bromine, and iodine are to be prepared. The reactions used to prepare the halogens will be carried out in water. The solvent hexane, which is not soluble in water, will be used to extract the halogens as they are formed. That is, the halogens are quite soluble in hexane, so if we have hexane in contact with the reaction mixture, the halogen formed will be captured in the hexane layer. Hexane is not part of any chemical reaction but just a convenient solvent for the halogens. Halogens are toxic and hexane is flammable, so work in the fume hood for all parts of this exercise. You may use approximate volumes of solutions. Hexane is an organic solvent and should not be poured down the drain. Place any used hexane in the container provided in the fume hood.

(a) **Preparing Chlorine** - Place 2 mL of bleach solution in a test tube. Add about 2 mL of hexane. The hexane floats on top of the water. Add about 2 mL of 1 M hydrochloric acid. Cork the tube and mix by repeatedly inverting the tube for a few minutes. Describe any changes and save the tube and its contents for part (b). The reaction that forms chlorine is: $HOCl + H^+ + Cl^- \longrightarrow Cl_2 + H_2O$
Why is it not a good idea to use household bleach with acidic cleaning solutions?

(b) Preparing Bromine - The chlorine formed in part a will be used to form bromine from bromide ion. You will need a clean test tube and an eye dropper. Remove the cork from the tube of part (a). Use your wash bottle to carefully add water to the tube to raise the level of the hexane so that it can be removed. Use an eye dropper to transfer the hexane layer to a clean test tube. Add 1 mL of 0.1 M KBr to the test tube. Cork the tube and mix by repeatedly inverting the tube for a few minutes. Describe any changes and give a balanced net-ionic equation for the formation of bromine. (See Table 29-1.) Save the tube and its contents for part (c).

(c) Preparing Iodine - The bromine formed in part b will be used to form iodine from iodide ion. You will need a clean test tube and an eye dropper. Remove the cork from the tube of part (b). Use your wash bottle to carefully add water to the tube to raise the level of the hexane so that it can be removed. Use an eye dropper to transfer the hexane layer to a clean test tube. Add 1 mL of 0.1 M KI to the test tube. Cork the tube and mix by repeatedly inverting the tube for a few minutes. Describe any changes and give a balanced net-ionic equation for the formation of iodine. (See Table 29-1.)

29 REPORT SHEET EXPERIMENT 29

Name _____

Lab Section _____   Due Date _____

1. Observing Redox Reactions

2. Some Typical Oxidation-Reduction Reactions

(a)

(b)

(c)

(d)

(e)

(f)

(g)

3. The Halogens and Halide Ions

Element	Symbol	Molecular Formula	Halide Ion Formula
Fluorine	F		
Chlorine	Cl		
Bromine	Br		
Iodine	I		

Element	Color of Element	Color of Hexane Solution
Chlorine		
Bromine		
Iodine		

Formula of Compound	Color and Form of Compound	Color of Water Solution
_____		
_____		
_____		
_____		

What do you conclude about the color of the halide ions?

4. Preparing Halogens

(a) Preparing Chlorine

(b) Preparing Bromine

(c) Preparing Iodine

29 QUESTIONS EXPERIMENT 29

Name_____

Lab Section _____ Due Date _____

1. Give definitions for the following terms:

(a) oxidation

(b) reduction

2. Using Tables 29-1 and 29-2, give the balanced net-ionic equations for any reactions that may occur when the following solutions are mixed.

(a) A $Na_2Cr_2O_7$ solution made acidic with sulfuric acid is added to a solution of potassium iodide, KI.

(b) A piece of zinc metal is added to hydrochloric acid.

(c) A piece of copper metal is added to a $Hg(NO_3)_2$ solution.

3. Using your observations and reactions in part 4 of this experiment, explain the following statement: Any halogen in the elemental form will displace any halide ion which is below it in the periodic table.

30 Equilibrium and Le Châtelier's Principle

Objective

The purposes of this exercise are to experimentally observe some equilibrium systems and test how changes can affect the equilibrium reactions.

Terms to Know

Chemical Equilibrium - A dynamic equilibrium between reactants and products in which the rate that the reactants form the products equals the rate that the products form the reactants. A state of chemical equilibrium can exists between reactants and products in a reversible chemical reaction.

Reversible Reaction - A chemical reaction in which the reactants can form the products or the products can form the reactants, depending on conditions.

Equilibrium Constant Expression - An expression, involving the ratio of the molar concentrations of the products to the reactants, which represents the fact that the ratio of concentrations is equal to a constant value for an equilibrium reaction. For the general reaction: $rR \rightleftharpoons pP$ the equilibrium constant expression is $K_{eq} = \dfrac{[P]^p}{[R]^r}$

Equilibrium Constant - The K_{eq} representing the constant numerical value corresponding to the ratio of the molar concentrations of the products to the reactants in an equilibrium reaction. The numerical value of an equilibrium constant is determined by substituting experimental values of equilibrium concentrations into the equilibrium constant expression for an equilibrium system.

Le Châtelier's Principle - When a factor affecting an equilibrium system is changed, the equilibrium will shift in a manner that tends to counteract the change.

Discussion

A reversible chemical reaction is a reaction in which the reactants can form the products and the products can form the reactants. In a reversible reaction, both the forward and reverse reactions occur and act in opposition. When a reaction of this type occurs, it soon reaches a state of dynamic equilibrium. Dynamic equilibrium is established when the rate of the forward reaction equals the rate of the reverse reaction. At equilibrium, the reactions occur continually but a balance of rates exists. Consequently, when a reversible reaction is at equilibrium, specific concentrations of reactants and products are present. An equilibrium is denoted by a double arrow between reactants and products.

The concentrations of gaseous species and dissolved species involved in a reversible reaction become constant when the reaction reaches equilibrium. This is true even though the reactions are continually occurring. The balance in concentrations of gaseous or dissolved species can be expressed in terms of an equilibrium constant. The constant comes from the fact that at equilibrium the ratio of the concentrations of the products to the concentrations of the reactants must equal some fixed and constant value at a given temperature. The equilibrium constant expression for a reaction that can be represented as

$$\text{reactants} \qquad\qquad \text{products}$$

$$2A \; \rightleftharpoons \; B$$

is written as

$$K_{eq} \;=\; \frac{[B]}{[A][A]} \;=\; \frac{[B]}{[A]^2}$$

where A and B are some gaseous or dissolved species, K_{eq} is the equilibrium constant and the square brackets [ ] represent concentrations in moles per liter. Note that, since species A is essentially two reactants, its concentration ends up to be squared in the expression. In general, the power used for any species in the equilibrium constant expression is its coefficient in the equation. Reactants or products in the form of solids or liquids are not included in an equilibrium constant expression.

A reaction system in equilibrium will remain in equilibrium indefinitely unless something upsets the equilibrium. At equilibrium, a balance of concentrations exists and, if a concentration is changed in any way, the equilibrium can be upset. Le Châtelier's principle is a useful guide to predict what may happen when an equilibrium is upset. A statement of the principle is: When an equilibrium system is upset by a change in any factor affecting the equilibrium, the equilibrium shifts in a direction that tends to counteract the change.

At equilibrium, a balance of concentration exists. If any one of the concentrations is increased by adding an additional amount of a species, the equilibrium will shift in a direction which tends to decrease the increase in concentration to keep the equilibrium balance. This is what the principle indicates. The idea is that the equilibrium shifts in one direction or the other to maintain the concentration balance. However, it is important to note that, in an equilibrium system involving a solid or a liquid, the concentration of the solid or liquid cannot be changed by adding more of the solid or liquid. On the other hand, the concentration of a gas or a dissolved species in an equilibrium system can be changed by adding or removing some of the gas or dissolved species.

An equilibrium system not only has a balance of concentrations of species, but also has an energy balance. Energy or heat is another factor that can affect an equilibrium system. In a sense, energy can be viewed as a reactant or product. If the energy supply is changed by heating or cooling, the equilibrium according to Le Châtelier's principle will shift. If an equilibrium system is heated, the equilibrium shifts in a direction that tends to use up the energy. The reaction shifts in the endothermic direction. If the system is cooled, the equilibrium shifts in a direction that tends to provide energy. The reaction shifts in the exothermic direction.

Laboratory Procedure

1. EQUILIBRIUM AND LE CHÂTELIER'S PRINCIPLE

Several equilibrium systems will be investigated and the effect of a temperature or concentration change will be observed. Record your observations.

(a) Saturated Sodium Chloride Solution.

Fill a large test tube with solid NaCl to a depth of about 3 cm. Add 10 mL of distilled water and gently heat the solution to boiling in a Bunsen flame. Stir the mixture with a glass rod. Set the solution aside to cool, go on with part (b), and return to this part later. Pour about 2 mL of the cooled solution into each of two small test tubes, and try not to transfer any solid salt. To the first tube add two or three drops of 12 M HCl or concentrated hydrochloric acid (H_3O^+ + Cl^-). (DANGER: Very concentrated acid.) To the second tube add a large crystal of rock salt. Describe any changes that occur in either tube.

The equilibrium in a saturated solution of sodium chloride is:

$$NaCl(s) \rightleftharpoons Na^+(aq) + Cl^-(aq)$$

Using this equilibrium and Le Châtelier's principle, explain your observations.

(b) Nitrogen Dioxide - Dinitrogen Tetroxide Equilibrium.

Nitrogen dioxide, NO_2, is a red-brown gas and dinitrogen tetroxide, N_2O_4, is a colorless gas. The two gases exist in equilibrium. Put some ice with water in a 400-mL beaker. Obtain a sealed tube of the equilibrium mixture from the stockroom and note the color of the mixture. Place the tube in the ice-water mixture for a few minutes. Go on to part (c) and return to this part later. Note any color change in the tube after it has cooled. Remove the tube from the ice water, dry it and, by gently rubbing with your hands, warm the tube. Note any color changes. Write an equilibrium equation for the reversible reaction involved in this section. Write the word "energy" on the side of the equation that correlates with your observations. Using Le Châtelier's principle, explain your observations.

(c) Saturated Ammonium Chloride Solution.

Fill a large test tube with solid NH_4Cl to a depth of about 3 cm. Add 5 mL of distilled water and stir the mixture with a glass rod. After mixing, carefully heat the tube in the Bunsen flame until it just begins to boil. Keep heating without boiling until all of the solid dissolves. Hold the test tube under a cold water tap and allow the solution to cool. Note the tube as it cools or after cooling. Compare this to what happened upon heating. Write an equilibrium equation for a saturated solution of ammonium chloride. (This is similar to the equation in part (a).) Write the word "energy" on the proper side of the equation and explain your observations using Le Châtelier's principle.

(d) Iron(III) Ion and Thiocyanate Ion in Equilibrium.

Fill two test tubes each with 10 mL of distilled water. Add 5 drops of 0.1 M $FeCl_3$ to one tube and 1 drop of 1 M NH_4CNS to the other. Pour one solution into the other and describe your results. The equilibrium involved is:

$$Fe^{3+}(aq) + CNS^-(aq) \rightleftharpoons FeCNS^{2+}(aq)$$

The $FeCNS^{2+}$ is red in color.

Pour one third of the solution into each of three test tubes. Keep one for color comparison. To one tube add several drops of 0.1 M $FeCl_3$ and describe any change.

To another tube add a few drops of 1 M NH_4CNS and describe any change. Use the above equilibrium and Le Châtelier's principle to explain any changes.

(e) Bromthymol Blue in Acid and Base Solutions.

Bromthymol blue is a complex organic compound which we will represent as HBB. In water it exists in the equilibrium

$$HBB(aq) + H_2O \rightleftharpoons H_3O^+(aq) + BB^-(aq)$$

The HBB has a yellow color and the BB^- ion has a blue color.

Add about 5 mL of distilled water to a large test tube followed by 6 drops of the bromthymol blue solution provided. Pour about 1 mL of a 6 M hydrochloric acid (H_3O^+ + Cl^-) solution into the tube. (DANGER: Acid.) Describe any change. Now, pour about 2 mL of a 6 M NaOH solution into the tube. (DANGER: Caustic.) Describe any change. The hydroxide ion in the NaOH reacts with hydronium ion by the reaction:

$$H_3O^+(aq) + OH^-(aq) \rightleftharpoons 2H_2O$$

This reaction decreases the concentration of hydronium ion. Use the above equilibrium reaction and Le Châtelier's principle to explain your observations.

(f) Cobalt Chloride solution

The equilibrium reaction involved in this section is

$$Co^{2+}(aq) + 4Cl^- \rightleftharpoons CoCl_4^{2-}$$

Place about 1 mL of a 0.1 M cobalt(II) chloride solution in a test tube. Note the color of the solution. Carefully add about 2 mL of 12 M hydrochloric acid as a source of chloride ion. (CAUTION: Very concentrated acid.) Note any color change. Using your observations label each cobalt species with its characteristic color. Write the color below the species in the equilibrium equation.

(g) Ammonia-Ammonium Ion In Equilibrium

In water, the ammonia and ammonium ion can exist in the following equilibrium.

$$NH_3 + H_2O \rightleftharpoons NH_4^+ + OH^-$$

Ammonia gas can turn red litmus blue when it is present in high enough concentration

232

above the equilibrium solution. Place about 2 mL sample of 0.1 M ammonium chloride solution in a test tube. Immerse a piece of red litmus paper into the air space above the solution in the tube. Do not allow the paper to touch the sides of the tube. Record the color of the paper. Carefully add about 2 mL of a 6 M NaOH solution into the tube. (CAUTION: Caustic.) Immerse a piece of red litmus paper into the air space above the solution in the tube and note the color of the paper. Do not allow the paper to touch the sides of the tube and be careful not to breathe any of the fumes coming from the tube. Use the above equilibrium reaction and Le Châtelier's principle to explain your observations.

2. EQUILIBRIUM EXPRESSIONS

Equilibrium expressions show an equilibrium constant, K_{eq}, as the ratio of the concentrations of the reactants. For the equilibrium reactions in parts (b), (d), (e), (f) and (g), write an appropriate equilibrium constant expression. Remember that only species in solution and gases are included in the expression. Solids and liquids are never included in the expression and water can be excluded from expressions involving equilibrium in water solutions.

(b)   (d)

(e)   (f)

(g)

30 REPORT SHEET EXPERIMENT 30

Name _____

Lab Section _____ Due Date _____

1. Equilibrium and Le Châtelier's Principle

(a) Saturated Sodium Chloride Solution

equilibrium equation:

(b) Nitrogen Dioxide - Dinitrogen Tetroxide Equilibrium

equilibrium equation:

(c) Saturated Ammonium Chloride Solution

equilibrium equation:

(d) Iron(III) Ion and Thiocyanate Ion in Equilibrium

equilibrium equation:

(e) Bromthymol Blue in Acid and Base Solutions

equilibrium equation:

(f) Cobalt Chloride Solution

equilibrium equation:

(g) Ammonia-Ammonium Ion Solution

equilibrium equation:

2. Equilibrium Expressions

(b)

(d)

(e)

(f)

(g)

30 QUESTIONS EXPERIMENT 30

Name _____

Lab Section _____ Due Date _____

1. Given the following chemical equilibrium,

$$CO_2(g) + H_2(g) \rightleftharpoons CO(g) + H_2O(g)$$

use Le Châtelier's principle to explain the following observations:

(a) Upon cooling, the equilibrium shifts towards the CO_2/H_2 side.

(b) A decrease in the concentration of CO causes the equilibrium to shift towards the CO/H_2O side.

2. Phenol red is a complex organic compound sometimes used as an indicator for testing swimming pool water. We can represent phenol red by the symbolic formula HPR. In water phenol red is involved in the following equilibrium reaction:

$$HPR(aq) + H_2O \rightleftharpoons H_3O^+(aq) + PR^-(aq)$$

HPR has a yellow color and PR^- has a red color.

(a) If a sample of water with a relatively low concentration of H_3O^+ is tested with phenol red, what color will result?

(b) If a sample of water is tested and turns somewhat orange in color, what does this indicate?

3. Write equilibrium constant expressions for

(a) The equilibrium reaction in question 1.

(b) The equilibrium reaction in question 2.

31 The Synthesis of Aspirin

Objective

The purpose of this experiment is to prepare some aspirin and determine the percent yield of the reaction.

Terms to Know

Aspirin - A medicinal drug used as an analgesic (pain reliever) and an antipyretic (fever reducer). Its chemical name is acetylsalicylic acid

Theoretical Yield - The mass of product that could be formed from reactants based upon stoichiometric calculations.

Actual Yield - The mass of product actually formed in a reaction.

Percent Yield - The ratio of the actual yield to the theoretical yield expressed as a percent.

Büchner Funnel - A specially designed funnel used for vacuum filtration.

Discussion

Aspirin or acetylsalicylic acid is the most common medicinal drug in use today. Aspirin is an analgesic or pain reliever and an antipyretic or fever reducer. Aspirin is often recommended for minor pain but it does have side effects when used excessively or by persons who are susceptible. Aspirin is known to cause dizziness, nausea, upset stomach, and bleeding of the stomach. Before using aspirin to treat children suffering from flu, a doctor or pharmacist should be consulted.

Aspirin can be made by reacting salicylic acid and acetic anhydride as shown by the following equation:

$$C_7H_6O_3 \ + \ C_4H_6O_3 \ \longrightarrow \ C_9H_8O_4 \ + \ HC_2H_3O_2$$

| salicylic acid | acetic anhydride | aspirin | acetic acid |

In this experiment, aspirin is prepared by reacting salicylic acid and acetic anhydride. Aspirin is quite insoluble in water so it can be crystallized in water and separated by filtration. Specific amounts of the reactants will be mixed to produce aspirin. The aspirin produced will be isolated and purified. However, it will not be pure enough for you to use.

Laboratory Procedure

CAUTION: Some of the chemicals used in this experiment are dangerous or very irritating to eyes and skin. Use them with care as instructed. Wear safety goggles and avoid breathing any fumes by keeping the flask away from your nose.

For the experiment you will need a 125-mL Erlenmeyer flask, a 400- mL beaker, two 100-mL beakers, a Büchner funnel (See Fig. 31-1), a filter flask and a one-holed rubber stopper to fit the 125-mL flask.

Obtain about 10.3 g of solid salicylic acid. Weigh a clean, dry 125-mL flask to 0.01 g. Transfer the sample of salicylic acid to the flask using a piece of creased paper. Weigh the flask plus the salicylic acid to 0.01 g. Subtract the masses to find the mass of salicylic acid used.

Fill a 400-mL beaker about two-thirds full of water and bring it to a boil by placing it on a wire gauze supported by a ring over a Bunsen burner flame. Take the 125-mL flask to the fume hood. Measure out about 13 mL of acetic anhydride in a graduated cylinder to the nearest 0.1 mL. (CAUTION: Acetic anhydride is very irritating. Do not breath the vapors. If you spill any on your skin, wash it off with large amounts of water.) Add the acetic anhydride to the flask. Keep the flask in the hood and add concentrated sulfuric acid from the dropper bottle provided. (CAUTION: Sulfuric acid is dangerous to the skin and eyes. If you spill or splash any on you, wash it off immediately using large amounts of water.) Carefully add 15 drops of concentrated sulfuric acid to the flask, swirling the flask to mix after the addition of each 2 to 3 drops of acid.

Place the one-hole stopper in the flask. Swirl to mix all of the chemicals. Place the flask in the beaker of gently boiling water. Control the boiling of the water by adjusting the burner flame. Heat the contents of the flask for 15 minutes. If any solid remains at this time, swirl the flask and continue heating for 10 minutes.

Remove the flask and cool by placing it in running tap water. Cool further by removing the stopper and adding about 30 mL of ice water and placing the flask in a beaker of crushed ice. Aspirin crystals should form at this time. If crystals are slow to appear, it may help to scratch the inside of the flask with a stirring rod. Leave the flask in the beaker of ice until it appears that no more crystals are forming.

Set up a Büchner funnel and filter flask as shown in Figure 31-1 (see the next page) and connect the flask to an aspirator. This apparatus will be used to filter the aspirin by suction filtration. Place a piece of filter paper in the funnel and wet the paper. Turn on the suction and filter the aspirin by transferring the contents of the flask to the filter. Use a spatula and a stream of water from your wash bottle to transfer the solid aspirin to the funnel. When the filtration is complete, turn off the suction and lift the filter paper with the aspirin out of the funnel. Transfer all of the aspirin to a 100-mL beaker and scrape any aspirin off of the filter paper with a spatula.

The aspirin is not pure since it will contain some of the reactants. It can be purified by recrystallizing it from alcohol. (CAUTION: Even after purification, the aspirin will not be pure enough to use so do not attempt to use it.)

Take the beaker of impure aspirin to the hood. Add about 25 mL of pure ethyl alcohol. (CAUTION: Ethyl alcohol is very flammable so do not work near a Bunsen flame. Make sure no burners are in the fume hood.) Stir the alcohol and aspirin mixture. If the crystals do not all dissolve, bring a beaker of hot water to the hood, hold the beaker in the water and stir the contents to dissolve the aspirin.

After all of the solid has dissolved, add about 70 mL of distilled water and place the beaker in a slightly larger beaker of ice. When crystallization again appears to be complete filter the aspirin by suction filtration. Suck as much liquid as possible from the aspirin before turning off the suction. Remove the filter paper with the aspirin and spread the product on a piece of paper towel. Dry the aspirin by placing the paper towel under a heat lamp or by letting it sit in your drawer for a few days.

To find the mass of the aspirin produced, weigh a small beaker to 0.01 g. Scrape the dry aspirin into the beaker and weigh it to 0.01 g. Subtract for the mass of aspirin.

From the equation for the reaction between salicylic acid and acetic anhydride and the masses of the reactants, calculate the expected amount of aspirin. To find the mass of acetic anhydride from the volume used you will need the density of acetic anhydride. Look up the density in a reference book or consult your instructor. This is a limiting reactant question so determine which of the reactants will form the lesser amount of aspirin. This will be the theoretical yield. The actual yield of aspirin will be less than the theoretical yield because of incomplete reactions and loss of product during filtration. The actual yield is the mass of aspirin actually formed in the experiment. The percent yield is the ratio of the actual yield to the theoretical yield expressed as a percent. Calculate the percent yield.

$$\% \text{ yield } = \frac{\text{actual yield}}{\text{theoretical yield}} \times 100$$

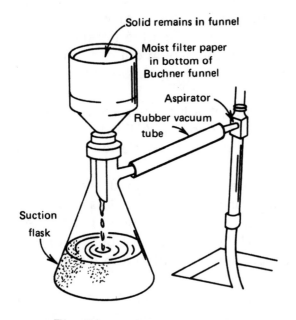

Fig. 31-1

31 REPORT SHEET EXPERIMENT 31

Name _____

Lab Section _____   Due Date _____

Preparation of Aspirin

Mass of flask and salicylic acid   _____

Mass of flask   _____

Mass of salicylic acid   _____

Volume of acetic anhydride   _____

Density of acetic anhydride   _____

Mass of acetic anhydride   _____

Mass beaker and aspirin   _____

Mass of beaker   _____

Mass of aspirin (actual yield)   _____

Theoretical yield (Show setup and results of calculations.)

Percent yield

31 QUESTIONS EXPERIMENT 31

Name _____

Lab Section _____   Due Date _____

1. Use the reaction of salicylic acid and acetic anhydride to determine the number of grams of aspirin which could be formed from 2.20 kg of salicylic acid.

2. Use the equation for the formation of aspirin from salicylic acid to answer the following question. A typical aspirin tablet contains 5 grains of aspirin. There are 15.4 grains per gram. How many grams of salicylic acid are needed to form 5 grains of aspirin?

3. About 40 billion pounds of aspirin are manufactured in the United States each year. If a typical aspirin tablet contains 5 grains of aspirin and there are 15.4 grains per gram, how many aspirin tablets could be made from 40 billion pounds of aspirin?

32 Soaps and Detergents

Objective

To observe some properties of soaps and detergents.

Terms to Know

Soaps - sodium-fatty acid ion compounds that have special cleansing properties.

Saponification - The chemical process by which fats are treated with sodium hydroxide solution to make soaps.

Hydrophilic - Water soluble or water liking as in the ionic end of a fatty acid ion.

Hydrophobic - Water repelling as in the hydrocarbon end of a fatty acid ion.

Detergent Action - The process by which soaps or synthetic detergents dissolve greasy materials and allow them to be suspended in water.

Syndets or Synthetic Detergents - Synthetic chemicals that display detergent action and are more useful than soaps in hard water.

Hard Water - Natural waters containing relatively high concentrations of calcium ion and magnesium ion.

Discussion

Soaps are chemicals having special cleansing properties. They are made from animal fats and vegetable oils. Fats and oils can react with concentrated solutions of sodium hydroxide to form the compound glycerol and organic ions called fatty acid carboxylate ions or simply fatty acid ions:

$$fat + OH^- \longrightarrow glycerol + 3\,RCOO^- \quad (RCOO^- \text{ a general symbol for carboxylate ions})$$

This type of reaction is called saponification which means soap making. The fatty acid ions can be precipitated out of solution by making the solution very concentrated in sodium chloride:

$$RCOO^- + Na^+ \longrightarrow RCOONa$$

The resulting sodium-fatty-acid-ion compound is soap. Since fats and oils contain a

243

variety of different fatty acids soaps are mixtures of various sodium-fattyacid-ion compounds. The formation of soap from palm oil, for example, is shown below.

saponification

palm oil $\xrightarrow{\hspace{3cm}}$ palmitate ion + oleate ion + other fatty acid ions + glycerol

$$\text{palmitate ion + oleate ion + others} \xrightarrow{\text{NaCl}} \text{sodium palmitate + sodium oleate + etc.}$$

A soap making company became famous using palm and olive oils to make soaps. What is the name of this company?

Soaps function as cleansing agents because of the unique structure of these fatty acid ions. When an object is dirty, it usually is a result of adhering layers of grease or oil containing dust and foreign particles. When the object is washed with water only, much of the "dirt" is not washed away since nonpolar greases and oils are not soluble in water. However, when soap is placed in the water, it dissolves to give fatty acid ions, as exemplified by the palmitate ion:

$$CH_3CH_2CH_2CH_2CH_2CH_2CH_2CH_2CH_2CH_2CH_2CH_2CH_2CH_2CH_2COO^-$$

long-chain hydrocarbon, hydrophobic end   ionic, hydrophilic end

The ionic end of such ions is very water-soluble (hydrophilic), but the long-chain hydrocarbon ends are hydrophobic and are not very water soluble. These long-chain hydrocarbon portions are responsible for the waxy feel of soap. When a soap is dissolved in water the hydrophobic ends of the fatty-acid ions accumulate on top of the water. This breaks down the surface tension of the water and gives soap solutions their characteristic sudsing or bubble making effect. The long-chain hydrocarbon ends of fatty-acid ions are soluble in nonpolar materials like oils and greases. The oil-attracting ends of the fatty-acid ions become embedded in the oil and grease layers, but the ionic ends of the fatty-acid ions remain dissolved in the water. This tends to pull the grease and oil particles into the solution as they become surrounded by a layer of soap ions. In this way the suspended grease and oil particles can be rinsed away. This kind of cleansing action is called detergent action.

When soaps are used in hard water containing relatively high concentrations of calcium and magnesium ion, soap scum can form, eliminating the soap and its cleansing action:

$$2\,RCOO^- + Ca^{2+} \text{ or } Mg^{2+} \longrightarrow (RCOO)_2Ca \text{ or } (RCOO)_2Mg \text{ (insoluble soap scum)}$$

Soaps do not work well in hard water, and when they are used in hard water, they leave soap scum on washed items.

Soaps are semisynthetic compounds. This means that they are synthesized from natural products. They do not occur naturally but can be easily made from natural fats and oils. Other organic compounds that have detergent action are synthesized from petroleum chemicals. Such compounds are called syndets (synthetic detergents) or simple detergents. Most syndets are sodium compounds of substituted benzene sulfonate, called linear alkyl sulfonates (LAS):

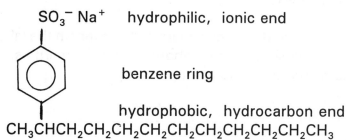

SO_3^- Na^+ hydrophilic, ionic end

benzene ring

hydrophobic, hydrocarbon end

$CH_3CHCH_2CH_2CH_2CH_2CH_2CH_2CH_2CH_2CH_2CH_3$

In LAS syndets the $-SO_3^-$ end of the ions are water-soluble, and the hydrocarbon ends are oil-attracting. Syndets act as cleansing agents in the same manner as described above for soaps. The advantage of syndets is that they do not form scum in hard water. In the United States synthetic detergents are more widely used than soaps. In this exercise you are going to make a sample of soap. Also the properties of soaps and syndets in various solutions will be observed.

LABORATORY PROCEDURE

1. MAKING SOAP

Set up a wire gauze on an iron ring supported by a ring stand. Adjust the ring so that the Bunsen burner flame reaches the bottom of the wire gauze. Place another iron ring slightly above the first ring. This second ring is a safety feature to keep the flask that you will be heating from accidently falling. Using a spatula about 5 grams of the fat supplied to a 125-mL flask. Push the fat sample to the bottom of the flask. A sample is on display to show how much to use. Add a few boiling chips to the flask. Use a paper towel to wipe any fat that adheres to the top of the flask. Now add about 20 drops of ethyl alcohol to the flask. Obtain about 2 mL of very concentrated (50 %) sodium hydroxide, NaOH, solution in a small beaker. (DANGER: The sodium hydroxide solution is very caustic and when it is heated it will be especially dangerous. YOU MUST WEAR SAFETY GOGGLES.) Add 10 drops of the NaOH solution to the flask.

Place the flask on the wire gauze and be sure to secure it with the second iron ring. Do not let the metal of the ring touch the flask. Light the Bunsen burner and heat the contents of the flask so that it foams and boils for a few minutes. After it stops foaming shut off the heat. (NOTE: If the fat starts to burn and turns black and smokes, turn off the heat immediately.) Now quickly add 20 more drops of the NaOH solution to the hot liquid and mix with your spatula. Allow the flask to cool. Remove the flask from the ring stand and place it on your desk. Obtain about 150 mL of saturated NaCl solution. Pour the salt solution into the flask. Mix with your spatula and as the soap floats to the top of the flask remove it and place it on a paper towel. To test the sudsing ability of your soap add 10 mL of deionized water to a test tube. Place a small amount of your soap into the test tube, stopper it and shake vigorously. Record your observations. For comparison, use your spatula to add a small amount of the fat to a test tube. Add 10 mL of deionized water to the tube, stopper it and shake vigorously. Record your observations.

2. TESTING SOAPS AND DETERGENTS

You will find some soap flakes and solid commercial detergent in the laboratory. Use small pieces of paper to obtain small amounts, about the size of marbles, of these two substances. Use these samples for the following tests.

a. Flame Test

Light your Bunsen burner. Pick up a small sample of soap flakes on the tip of a spatula and hold it in the flame. Describe the flame. Let the spatula cool and, then, clean it. Pick up a small amount of detergent on the tip of the spatula and hold it in the flame. Describe the flame. A characteristic of compounds containing benzene is that they give a smoky black flame upon burning.

b. Soap and Detergent Solutions in Acids and Salt Water

Place the soap flakes in a small test tube and the detergent sample in another small test tube. Fill the tubes with deionized water, stopper each tube and shake vigorously to make solutions. Label the tubes to distinguish the soap from the detergent. These solutions are to be used in this part and in part 3. Use litmus paper to test each of the solutions. Record your results.

Pour 5 mL samples of deionized water into two large test tubes. Add 5 drops of your soap solution to one tube and 5 drops of your detergent solution to the other tube. Stopper the tubes and shake vigorously. Now pour 5 mL samples of dilute hydrochloric acid (DANGER: 6 M HCl) into each tube, stopper the tubes and shake vigorously. Record your results.

Pour 5 mL samples of saturated sodium chloride solution into two large test tubes. Add 5 drops of your soap solution to one tube and 5 drops of your detergent solution to the other tube. Stopper the tubes and shake vigorously. Record your observations.

3. SOAPS AND DETERGENTS IN HARD WATER

You will need 6 test tubes. Be sure that they are clean. You need a stopper or cork for each tube. Mark the tubes A, B, C, D, E and F for reference. Add 5 mL of deionized water to tubes A, B, C and D and 5 mL of the hard water sample to tubes E and F. The hard water is available in the laboratory in an appropriately labeled bottle.

Add three drops of your soap solution to tubes A, B and E.

Add three drops of your detergent solution to tubes C, D and F.

Add two drops of vegetable oil to tubes B and C.

Stopper each tube tightly and shake each vigorously. Record your observations.

4. CLEANSING ACTION OF DETERGENT

Obtain a small amount of powdered detergent or liquid detergent in a small beaker. Remove all glassware from your desk. Add small amounts of tap water to all of your glassware and a bit of detergent to each vessel. Use a brush to thoroughly scrub the inside and outside walls of each vessel. Thoroughly rinse each vessel and dry if necessary. A glass vessel is clean when water does not bead or form droplets on its surface. Check your cleaned glassware and record your observations.

32 REPORT SHEET EXPERIMENT 32

Name _____

Lab Section _____ Due Date _____

1. MAKING SOAP

2. TESTING SOAPS AND DETERGENTS

a. Flame Test

b. Soap and Detergent Solutions in Acids and Salt Water

3. SOAPS AND DETERGENTS IN HARD WATER

Tube	Soap	Detergent
A Deionized Water		X
B Deionized Water + Oil		X
C Deionized Water + Oil	X	
D Deionized Water	X	
E Hard Water		X
F Hard Water	X	

4. CLEANSING ACTION OF DETERGENT

32 QUESTIONS EXPERIMENT 32

Name _____

Lab Section _____ Due Date _____

1. What is saponification?

2. Why are soaps considered to be semisynthetic compounds?

3. Why do soaps not work well in hard water?

4. Why is it not a good idea to use typical soaps in ocean water?

33 Biochemistry and Eggs

Objective

To observe some properties of proteins and lipids.

Terms to Know

Coagulation - A process in which liquid or semisolid material changes to a solid material.

Fats or lipids - Triglycerides or esters of the trihydroxy alcohol glycerol and long-chain carboxylic acids called fatty acids. Animal fats and vegetable oils are typical simple lipids.

Proteins - Biological polymers consisting of chains of hundreds or thousands of amino acids. Proteins are synthesized in cells and serve as structural units of cells, skin, muscles, bone interior, and nerves, and as enzymes and hormones. Proteins are an important food source for humans since they provide amino acids needed by the body to synthesize proteins.

Discussion

The three most important types of biochemicals are proteins, carbohydrates and lipids (commonly called fats). Carbohydrates and fats serve as energy sources but they have many other important uses in the body. Proteins have many structural functions in the body and are involved in numerous cellular functions. In this exercise you are going to observe some properties of proteins and lipids.

Chicken eggs are embryos that contain some interesting biochemicals. The shell of an egg is about 95% calcium carbonate and 5% protein by mass. Not including the shell, a typical egg contains about 12% protein, 9% fat (lipids), 2% carbohydrates and 0.5% cholesterol. Most of the fats and cholesterol are contained in the yolk along with some protein. The white of the egg is called the albumen. The white is mainly water and contains about 10% protein along with small amounts of minerals, glucose and fats. The major protein in egg white is called egg albumin.

Laboratory Procedure

Set up a water bath by placing a 400-mL beaker of water on a wire gauze supported by an iron ring attached to a ring stand. Heat the water with a Bunsen burner.

1 Decomposing Protein with a Strong Base (How to dissolve hairy grease balls.)

Use a spatula to place a scoop of Draino crystals in a large test tube. Use a rod to push a lock of your hair into the tube. Your hair is a convenient sample of protein. Working in the fume hood place about 5 mL of deionized water in the tube. Observe the tube and describe what happens. Let the tube sit in the hood for fifteen minutes. Mix the contents of the tube and describe the appearance of the hair. Carefully pour the solution down the drain in the hood and rinse the tube with water. Take care not to spill or splash the solution.

2 Coagulating the Protein Casein from Milk (Preparing Ms. Muffet's meal.)

Cows milk contains proteins (mainly the protein casein), lipids (in the form of butter fat), and carbohydrates (mainly in the form of lactose or milk sugar). Since butter fat is not soluble in water the fat portion of fresh milk floats on top of the water in milk. Homogenized milk is prepared by passing fresh milk through a very fine sieve so that small butter fat globules are formed and distribute in the water portion of milk in a "homogeneous" manner. The fat in homogenized milk does not float to the surface but stays suspended. Non fat milk is made by skimming the butter fat from fresh milk. This is why it is called skim milk.

Pour about 20 mL of non fat milk into a test tube. Test the milk with red and blue litmus paper and record your results. When milk is made more acidic the suspended protein in milk (called casein) will coagulate into an insoluble solid. This is what happens when lactobacillus is cultured in milk so that it forms lactic acid which, in turn, coagulates casein to form yogurt. Heat the tube of milk in your boiling water bath for about 5 minutes. Add 20 drops of 6 M acetic acid, $HC_2H_3O_2$. (DANGER:Acid) Stir the milk with a stirring rod to help coagulate the casein into curds. The liquid left after the casein coagulates is called the whey. So you have made some curds and whey. Describe the appearance of the contents of the tube. Use your rod or spatula to remove some of the coagulated solid casein and place it on a paper towel. Describe the solid and save it for part 5.

3 Coagulating Egg Albumin Using Heat (Preparing Egg Flower Soup)

Chicken egg whites contain a protein called egg albumin. The egg yolk contains some lipids as well as many other chemicals. Proteins can be coagulated by heat. Proteins have complex molecules held in a three-dimensional shape by delicate bonds. Heating can break these delicate bonds causing the protein to coagulate. When it coagulates the protein is said to be denatured. This is why milk will curdle when it is boiled and why egg white turns to a semi-solid material when it is cooked.

Obtain a fresh egg. Crack the egg and separate the yolk from the egg white. Place the egg white in one beaker and the yolk in another beaker (save it for parts 6 and 7). Place 20 mL of water in a test tube. Heat the water to boiling. Pour some of the egg white into the boiling water and stir with a stirring rod. Describe your results. Use a rod or spatula to remove some of the coagulated solid from the tube and place it on a paper towel. Describe the solid and save it for part 5. Save the extra egg white for parts 4 and 8.

4 Egg White Coagulation (Making Egg White White)

The protein in egg white can be denatured or coagulated using chemicals rather than heat. Heavy metal compounds including lead and mercury can disrupt protein molecules. In fact heavy metal poisoning can disrupt important body proteins. Strongly acidic solutions can denature protein molecules. This is one reason why acids can burn your skin. Alcohols can replace water in proteins and denature them. This is why some alcohols are used as mild disinfectants since they can denature the proteins of bacteria.

Place small samples of egg white in three wells of a spot plate. Add a drop of mercuric nitrate solution to one well, a drop of concentrated hydrochloric acid to another well and a drop of ethyl alcohol to the third well. Describe your observations.

5 Chemical Spot Test for Proteins (Out Damn Spot)

A simple spot test for proteins involves adding nitric acid to a sample to be tested. Nitric acid readily reacts with certain amino acids in common proteins to give a distinct deep yellow color. Obtain a spot plate. Place a small sample of casein (from part 2) into one spot, place a small sample of egg albumin (from part 3) into a second spot and a few finger nails (cut your own) into a third spot. Place 4 drops of concentrated nitric acid, HNO_3, into each spot. (**Danger:** Be very careful with nitric acid. If you drop it on your skin it will form a yellow spot indicating that your skin contains protein.) Record the appearance of the spots after 5 minutes.

6 Lipids in Water (Oil on Troubled Waters)

Lipids are typically nonpolar organic compounds. As nonpolar compounds they are not expected to be very soluble in a polar solvent like water. This is why when you cook fat containing animal cells many of them burst forming fat that floats on top of any water present.

Place about 15 mL of deionized water in a test tube. Using a dropper bottle of vegetable oil add 20 or 30 drops of oil to the water. Describe the appearance of the liquids in the tube. Stopper the test tube and shake the contents. Describe the contents of the tube. Egg yolk contains lipids suspended in water. Certain chemicals in the yolk act as agents that allow the water insoluble lipids to be suspended in water as an emulsion. These chemicals are called emulsifiers. Using an eye dropper add several drops of egg yolk to your test tube. Stopper the tube and shake it vigorously. Describe the results. Egg yolk emulsifiers are used to make an emulsion of vegetable oil in water known commonly as mayonnaise.

7 Extracting Lipids from Egg Yolk (Making a Greasy Spot)

Since lipids are nonpolar compounds they are readily soluble in nonpolar solvents. Pour the egg yolk from part 3 into a large test tube. Add 2 mL of hexane to the tube. Stopper the tube and mix the yolk with the hexane for a few minutes by continually inverting the tube. Set the tube down to let the hexane settle to the top. Take a 6 inch length of paper towel to the fume hood that contains the bottle of hexane. Lay the paper down and drop a few drops of hexane on the paper. Note that in a short time the hexane evaporates. Take your yolk-hexane tube to the hood. Pour a small amount of the upper hexane layer into a small beaker. Pour the hexane extract onto the paper towel and let it sit for a while. Describe the appearance after the hexane evaporates. The greasy spot contains some lipids extracted from the egg yolk.

8 Burning Egg Whites (Not over easy)

Proteins are polymers of amino acids. Proteins can be decomposed by combustion to give carbon dioxide, water and nitrogen compounds. When proteins are "burned" in the body by metabolism carbon dioxide, water and nitrogen compounds like urea are formed. In this exercise you are going to burn some egg white to decompose the protein and form some nitrogen containing compounds. When the burned material is made basic and heated some ammonia and ammonia-like compounds will be evolved. Their evolution can be detected with litmus paper.

Work in the fume hood. Pour your egg white sample from part 3 into a large test tube. Light a Bunsen burner. Hold the tube of egg white with a test tube holder making sure that the holder is near the top of the tube. Heat the egg white in the flame until it burns and turns black. Allow the tube to cool and add 1 mL of 6 M NaOH solution. (DANGER: Boiling NaOH solution is very dangerous. Be sure that you are wearing goggles.) Obtain a piece of red litmus paper and wet it with deionized water. Have a pair of tweezers ready to hold the paper in the mouth of the test tube. In the hood heat the test tube until the solution boils. Hold the wet litmus paper with the tweezers and carefully hold it in the mouth of the test tube. Do not let the paper touch the sides of the tube. Describe your observations.

9 Comparing Vegetable Oils (Monounsaturated versus Polyunsaturated)

At room temperature animals fats (butter, lard, tallow) tend to be semi-solids and vegetable oils tend to be liquids. The difference is related to the content of polyunsaturated fatty acids in the lipids. Saturated fatty acids make lipids semi-solids. This is why animal fats and products which contain partially hydrogenated vegetable oils (which means partially saturated), like margarine and Crisco, are semi-solids. The greater the percentage of polyunsaturated fatty acid components of vegetable oils the lower the freezing point of the oil.

Place some ice in a 250 mL beaker. Add a large sample of solid ammonium sulfate to the beaker and add some water to make a low temperature ice bath. Stir the bath until it gets very cold. If the ice melts add more as needed. It is important to have a very cold ice bath. Pour about 1 mL samples of the various oils on display into small test tubes. Label the tubes. Place these tubes in your ice bath. Allow the tubes to sit in the bath for a several minutes and describe any change in appearance. Which oil do you think contains the highest percentage of polyunsaturated fatty acid components. Which contains the highest percentage of saturated fatty acid components?

33 REPORT SHEET EXPERIMENT 33

Name _____

Lab Section _____ Due Date _____

1 Decomposing Protein with a Strong Base (How to dissolve hairy grease balls.)

2 Coagulating the Protein Casein from Milk (Preparing Ms. Muffets meal.)

3   Coagulating Egg Albumin Using Heat (Preparing Egg Flower Soup)

4 Egg White Coagulation (Making Egg White White)

5 Chemical Spot Test for Proteins (Out Damn Spot)

6 Lipids in Water (Oil on Troubled Waters)

7 Extracting Lipids from Egg Yolk (Making a Greasy Spot)

8 Burning Egg Whites (Not over easy)

9 Comparing Vegetable Oils (Monounsaturated versus Polyunsaturated)

33 QUESTIONS EXPERIMENT 33

Name _____

Lab Section _____ Due Date _____

1. What is the approximate percent of water in a typical egg?

2. Why is drain cleaner potentially dangerous to humans? Why is it not recommended that drain cleaner be added to a system that includes a septic tank?

3. List four ways that could be used to denature or coagulate proteins.

4. If you had a sample of vinegar and oil dressing and you added some egg yolk, what do you predict would happen when you shake the mixture vigorously?

5. Why does olive oil freeze more readily than other vegetable oils?

Appendix 1 Laboratory Techniques and Measurements

Terms to Know

Interpolation - Reading a calibrated scale by estimating the measured position as it occurs between the smallest calibration lines on the scale.

Bunsen Burner - A device used as a source of heat in which natural gas and air are mixed.

Metric Rule - A ruler calibrated in metric units of length. Metric rules are typically calibrated so that they can be read to the nearest 0.1 centimeter or 1 millimeter.

Graduated Cylinder - A calibrated vessel made of cylindrical glass used to measure the volumes of liquids.

Fire Polishing - Heating the ends of a newly cut piece of glass tubing so that the sharp edges melt.

Measurement and Measuring Instruments

Since chemistry is an experimental science, observations and measurements are very important. Special measuring devices are used to measure the properties of objects. These measuring devices or instruments are typically calibrated to be read in the appropriate metric unit. For example, length measurements are made using a calibrated ruler called a meter stick. Mass measurements are carried out using a balance which compares the mass of an object to calibrated masses called weights. Temperature measurement is made through the use of a thermometer.

A measuring instrument is calibrated to be read to a certain number of digits. Usually, a measuring instrument should be read to as many digits as possible. This sometimes requires estimating between the smallest calibrated divisions on the instrument. Such an estimation process is called interpolation and is described in Figure A1-1. Whenever you make a measurement, it is very important to always read the instrument as carefully as possible and to read the proper number of digits. In experiments you will be instructed to make measurements to specific numbers of digits. When you make a measurement, always record it as both a number and a unit (e.g., 2.3 cm, 5.78 g, 25.0 mL).

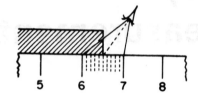

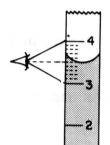

Fig. A1-1 Interpolation. The distance between divisions is visually divided into 10 equal parts and the visualized line closest to the object edge, mark, or bottom of the meniscus is noted. The distance between divisions may be divided into fewer parts depending on the calibration of the measuring device.

1. BALANCES AND WEIGHING

Various kinds of balances are used to measure the masses of objects. Your instructor will demonstrate the use of the kind of balance available in the laboratory.

The general rules for weighing materials are:

1. Be sure that the balance has been zeroed properly before using it.

2. Never place chemicals directly on the balance pan. Always use weighing paper or some container such as a small beaker.

3. Find out the maximum mass you can weigh on a balance and do not try to weigh objects that exceed the maximum mass.

4. Treat balances with much care. If you have trouble using a balance, ask for help.

5. Never weigh a hot object. Allow a heated object to cool to room temperature before you weigh it.

6. Always clean up any chemicals that you spill on the balance or in the weighing area.

WEIGHING BY DIFFERENCE   Typically the mass of a sample of material is found as follows:

1. Weigh an empty container or a piece of weighing paper.

2. Add the sample to the container or paper and weigh the combination using the same balance as used for the first weighing.

3. Subtract the mass of the container or paper from the mass of the combination to obtain the mass of the sample.

2. THE BUNSEN BURNER

A Bunsen burner mixes natural gas with air so that the gas can be burned and serve as a source of heat for laboratory experiments. Typical Bunsen burners are shown in Figure A1-2. The gas and air are mixed in the barrel of the burner. Note that there is an adjustable air intake vent on the barrel. The flow of gas is regulated by the gas valve connected to the gas pipe or, on some burners, there is a gas-regulating valve connected to the burner. To light the burner, turn on the gas valve to about half pressure (not full pressure). Hold a lighted match at the edge of the top of the barrel, but not directly over the top of the barrel. If a flint striker is used, hold the striker over the top of the barrel and strike it.

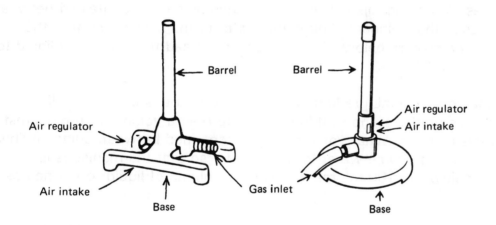

Fig. A1-2 Typical Bunsen Burners

Once the burner has been lit, the flame can be regulated by adjusting the gas flow and the air intake vent. The flame on a properly adjusted burner will have the appearance of a pale blue inner cone and a pale violet outer cone. The hottest part of the flame is at the tip of the inner blue cone. If there is too much gas in the flame, it will have a yellowish appearance. To obtain a proper flame, open the air intake. Too much air and gas pressure will form a flame that separates from the burner top. To adjust such a flame, lower the gas pressure and decrease the amount of air. If you want a flame that is cooler than the normal flame, decrease the amount of air by adjusting the air intake. A low burning flame can be obtained by adjusting the gas pressure using the gas valve.

PRACTICE EXERCISE: Light your Bunsen burner and adjust the flame to give a pale blue inner cone and a pale violet outer cone. Sketch a side view of the burner showing the flame and indicate the colors of the cones.

3. VOLUME MEASUREMENT

Careful volume measurements of liquids are carried out by use of graduated cylinders which are precisely calibrated cylindrical glass or plastic vessels.

PRACTICE EXERCISE: (a) Examine your 10-mL graduated cylinder. Note the large and small calibration lines on the cylinder. It is possible to read a volume to the nearest 0.1 mL in this cylinder. What volume can be contained between adjacent lines in the cylinder? Add some water to the cylinder so that the meniscus is somewhere below the 10 mL mark. Read the volume of liquid to the nearest 0.1 mL.

(b) Examine your 100-mL graduated cylinder. Note the large and small calibration lines on the cylinder. It is possible to read a volume to the nearest 0.5 mL in this cylinder. What volume can be contained between adjacent lines in the cylinder? Add some water to the cylinder so that the meniscus is somewhere below the 100 mL mark. Read the volume of liquid to the nearest 0.1 mL.

(c) Fill a small test tube with water and pour the water into your 10-mL graduated cylinder to measure the volume of the tube. Record the volume.

Volume of tube _____

How full would you have to fill this tube so that you would have about 2 mL of water? Add about 2 mL of water to the tube and pour it into the graduated cylinder to see how close you estimated. Now add about 5 mL of water to the tube and pour it into the graduated cylinder. How close did you estimate the volume? Sometimes you are instructed to measure about 1 mL, 2 mL or 5 mL of liquid volume. Often you can do this by estimating the volume in a clean test tube. If you need a precise volume measurement, use a graduated cylinder.

4. DISPENSING CHEMICALS

Chemicals that are used in the laboratory are sometimes called reagents. When pouring liquid chemicals from glass-stoppered reagent bottles, remove the stopper and hold it in your fingers while carefully pouring the liquid into the desired container. This technique is illustrated in Fig. A1-3. Never set a glass stopper on the desk. Hold it in your fingers. When pouring from a screw cap bottle, set the cap down on the top so that it does not become contaminated. Be sure to put the correct cap on the bottle after you have used it. If you spill any liquid or drip some on the side of the bottle, clean it up. Never pour excess chemicals back into the reagent bottles since this can contaminate the reagents.

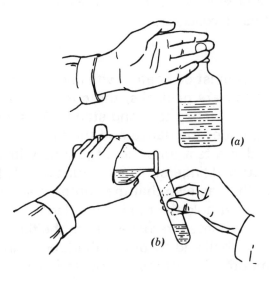

Fig. A1-3 Proper technique for pouring
liquids from a glass-stoppered bottle.

When obtaining samples of powdered or crystalline solids from a jar, pour the desired amount of the solid onto a small piece of clean paper or into a clean, dry beaker. To pour a solid, do not turn the jar upside down and dump. Carefully tilt the jar and rotate it back and forth to work the solid up to the lip. Then, using the same back and forth rotation, allow the desired amount of solid to fall from the jar. This is illustrated in Fig. A1-4. Be careful when transferring a solid from a jar. If you take too much, leave the excess on a piece of paper for other students or throw the excess away. Never put any solid back into the jar. Furthermore, never put wooden splints, spatulas, or paper into a jar of solid unless your instructor indicates that this is permissible. Solids may be transferred by using a piece of paper that has been creased down the center. See Fig. A1-4

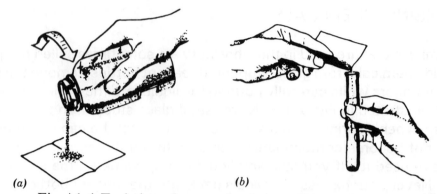

(a)   (b)

Fig A1-4 Technique for transferring a solid. (a) Place the solid on a creased piece of paper. (b) Slowly rotate the jar until the desired amount of solid falls out. (c) Using the creased paper, pour the solid into a container.

5. HEATING CHEMICALS

Solid and liquid chemicals must be heated with care to prevent explosions and accidents. To heat liquids in beakers or flasks, the container are placed on a wire gauze supported on an iron ring attached to a ring stand. As shown in Fig. A1-5 on page 263 the burner is placed under the gauze and the ring is adjusted for efficient heating. When heating a liquid in a test tube, hold the tube with a test tube holder. The test tube holder prevents burned fingers. The test tube should be carefully heated by moving it back and forth in the flame so that the contents are evenly heated. See Fig. A1-6 below. (DANGER: Never hold the tube in the flame without moving it back and forth.) As soon as the tube is heated, remove it from the flame. Failure to do this may cause the liquid to suddenly boil and fly out of the tube. Since this can be very dangerous, never point the mouth of the tube at yourself or your neighbor (or your instructor).

Fig. A1-6 Proper technique for heating a test tube of a liquid.

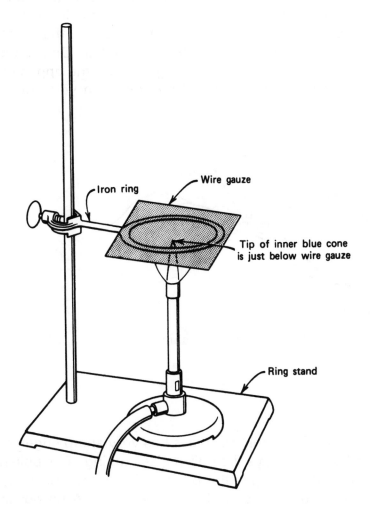

Fig. A1-5 Apparatus for heating liquids. The flask or beaker is placed on the wire gauze for heating with the Bunsen burner.

7. WORKING WITH GLASS

When setting up equipment for experiments, it is sometimes necessary to use glass tubing as glass connecting tubes or for such things as stirring rods. It is important to learn how to work with glass tubing. Always wear eye protection when working glass.

CUTTING GLASS TUBING   Wear eye protection when cutting glass. Obtain a length of glass tubing. Using a triangular file, a glass knife or a glass cutting wheel, make a scratch on the tubing at the desired point. Make a small scratch and do not extend it all the way around the tubing. If using a file, make the scratch with an outward motion of the file. Do not saw the tubing with the file. Once scratched, pick the tubing up and hold it so that the thumbs are together on the opposite side of the scratch, and then gently push the glass away from you with your thumbs. This is shown in Fig. A1-7.

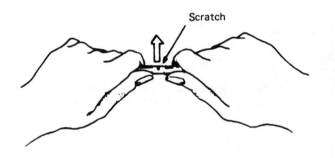

Fig. A1-7 Cutting glass tubing

FIRE POLISHING GLASS Freshly cut glass has very sharp edges that must be smoothed in the Bunsen flame. Always fire polish glass tubing. Fire polishing is done by rotating the end of the glass tubing in the Bunsen flame until the sharp edges are smooth. If the tube is held in the flame too long, it will seal. After polishing, a tube should be placed on a ceramic sheet or wire gauze to cool. Be careful of hot tubing. Do not touch the tubing until it is cool. Fire polish all ends of newly cut glass tubing.

SEALING TUBES AND MAKING STIRRING RODS A stirring rod can be constructed from glass tubing or from solid glass rod. Glass rod makes the best stirring rods. If rod is used, cut the desired length and polish each end. If tubing is used, cut a piece about 5 inches longer than the desired length of the rod. Hold the tube with both hands and place it at the tip of the Bunsen flame about 2 inches from one end of the tubing. Rotate or roll the tube in your fingers as it heats. When the tubing begins to melt at the point of heating, remove it from the flame and gently pull both ends to stretch the tubing. As the tubing stretches it will decrease in diameter. Set the tube aside to cool.

After cooling, use a file to cut the tube at the constriction. Place the constricted end of the tube in the hot part of the Bunsen flame and rotate the tube until the end melts and seals. Make sure to heat long enough to seal any small holes in the tip. After cooling, the other end of the tube can be sealed in a similar fashion.

BENDING GLASS TUBING Glass tubing can be bent to desired angles for use in glass delivery tubes and glass connecting tubes. When bending glass, always start with a length of glass that is longer than needed. This allows for easy heating without burning your fingers. After the bent glass has cooled, it can be cut to the desired length and fire polished. Attach a wing top flame spreader to the Bunsen burner and light the burner. Hold the tubing in the flame where the bend is desired and slowly rotate the tubing to heat evenly. (See Fig. A1-8.) It is important to continuously rotate the tubing while heating and to heat it over the entire section to be bent. When the glass is soft and begins to droop, remove it from the flame and bend the two ends in an upward direction to form the desired angle (see Fig. A1-8), and then set it on the ceramic sheet or wire gauze to cool.

INSERTING GLASS TUBING IN RUBBER STOPPERS The insertion of glass tubing into rubber stoppers can be dangerous if the glass breaks. First of all, make sure that all tubing is fire polished and cool. Lubricate the end of the tubing and the edges of the hole in the stopper with water or glycerol. Your hands can be protected with gloves or with a towel if desired. Place the tubing in the hole and work it in by twisting the tube or the stopper. Never push the tubing into the hole. Always twist the tubing by holding the portion of it that is closest to the stopper. Always be careful and do not hurry. If you have trouble, tell your instructor. If the tubing breaks and you cut yourself, tell your instructor immediately.

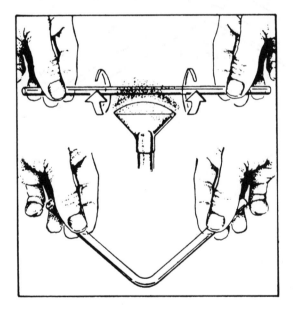

Fig. A1-8 Bending glass tubing

PRACTICE EXERCISE: Obtain some glass tubing and make the glass objects shown in Figure A1-9. Note that the glass items in the figure are not shown to scale so you will need a metric rule to estimate lengths. For the angular and short tubes, work with much longer lengths of tubing than you need and then cut the tubing to the desired lengths after bending or sealing. Do not forget to fire polish after cutting tubing. Furthermore, be careful not to touch the hot glass. After making your glass objects, obtain your instructor's approval.

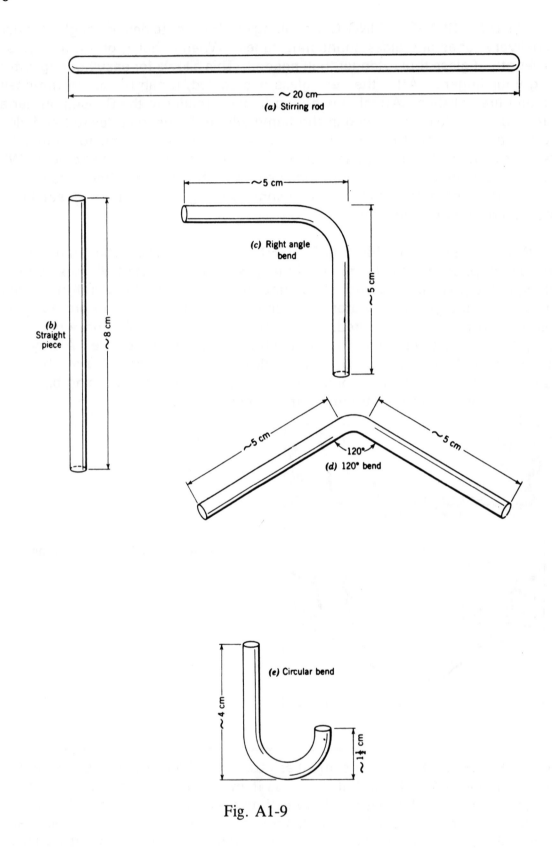

(a) Stirring rod

(b) Straight piece

(c) Right angle bend

(d) 120° bend

(e) Circular bend

Fig. A1-9

Appendix 2
Vapor Pressure of Water

Temperature °C	Vapor Pressure Torr	Temperature °C	Vapor Pressure Torr
0	4.6	25	23.8
5	6.5	26	25.2
10	9.2	27	26.7
15	12.8	28	28.3
16	13.6	29	30.0
17	14.5	30	31.8
18	15.5	40	55.3
19	16.5	50	92.5
20	17.5	60	149.4
21	18.6	70	233.7
22	19.8	80	355.1
23	21.2	90	525.8
24	22.4	100	760.0

Appendix 3
Laboratory Acids and Bases

Solution	Label Formula	Molarity
Concentrated Sulfuric Acid	conc. H_2SO_4	18 M H_2SO_4
Concentrated Hydrochloric Acid	conc. HCl	12 M HCl
Concentrated Nitric Acid	conc. HNO_3	16 M HNO_3
Concentrated Acetic Acid Glacial Acetic Acid	conc. $HC_2H_3O_2$	17 M $HC_2H_3O_2$
Concentrated Phosphoric Acid	conc. H_3PO_4	15 M H_3PO_4
Concentrated Ammonia (Sometimes labeled as ammonium hydroxide.)	conc. NH_3	15 M NH_3
Dilute Sulfuric Acid	dil. H_2SO_4	3 M H_2SO_4
Dilute Hydrochloric Acid	dil. HCl	6 M HCl
Dilute Nitric Acid	dil. HNO_3	6 M HNO_3
Dilute Acetic Acid	dil. $HC_2H_3O_2$	6 M $HC_2H_3O_2$
Dilute Ammonia (Sometimes labeled as ammonium hydroxide.)	dil. NH_3	6 M NH_3
Dilute Sodium Hydroxide	dil. NaOH	6 M NaOH